THE NETWORK ALWAYS WINS

How to Influence Customers, Stay Relevant, and Transform Your Organization to Move Faster than the Market

PETER HINSSEN

Mc
Graw
Hill
Education

New York   Chicago   San Francisco   Athens   London
Madrid   Mexico City   Milan   New Delhi
Singapore   Sydney   Toronto

1 2 3 4 5 6 7 8 9 0   DOC/DOC   1 2 1 0 9 8 7 6 5

ISBN   978-0-07-184871-8
MHID   0-07-184871-1

e-ISBN   978-0-07-184870-1
e-MHID   0-07-184870-3

Library of Congress Cataloging-in-Publication Data

Hinssen, Peter
    The network always wins : how to influence customers, stay relevant, and transform your organization to move faster than the market / Peter Hinssen.
        pages cm
    ISBN 978-0-07-184871-8 (hardback)
    ISBN 0-07-184871-1 (hardback)
    1. Strategic planning.   2. Technological innovations—Management.   3. Information technology—Management.   4. Management.   I. Title.
    HD30.28.H5584 2015
    658.4'012—dc23   2015001221

McGraw-Hill Education books are available at special quantity discounts to use as premiums and sales promotions or for use in corporate training programs. To contact a representative, please visit the Contact Us pages at www.mhprofessional.com.

All illustrations are by Vera Ponnet of Saflot.

TO MY CHILDREN

———————————

The most amazing wonder of the network between Valentine and me.
The eyes through which I gaze at the world.
They have taught me more than any book I have ever read.

Aida and Loren, this book is for you.

CONTENTS

THE ERA OF NETWORKED HEALTH

CHAPTER 6 WHEN ORGANIZATIONS BECOME NETWORKS OF INNOVATION

CHAPTER 7 CREATION AND DESTRUCTION

CHAPTER 8 STRATEGY FOR THE AGE OF NETWORKS

ACKNOWLEDGMENTS

I would like to thank all the people who were vital in making this book a reality: Ilse De Bondt, José Delameilleure, Nadia Del Rio, Mary Glenn, David Hayward, Ciel Jolley, Rick Judge, Marc Lerouge, Jim Lubinskas, Devanand Madhukar, Peter McCurdy, Luc Osselaer, Taunya Renson-Martin, Chantal Van de Ginste, Laurence Van Elegem, and Marianne Vermeulen.

Thanks to all the people who, directly or indirectly, have been an inspiration for me for this book: Jamie Anderson, Rob Goffee, Costas Markides, Michael Nowlis, Steven Van Belleghem, Sean Gourley, Mark Zawacki, Glenn Morgan, Mike McNamara, Bart De Crem, Wim De Waele, Bart Van Hooland, André Duval, Alex Brabers, Davy Kestens, Yoeri Roels, Gary Hamel, Thomas Leysen, Whitney Bouck, Walter De Brouwer, Luc Verhelst, Peter Vanderauwera, Tom Standage, Sofie Bruynooghe, Thierry Geerts, Peter Strickx, Steven De Smet, Steve Van Wyck, Kees Smaling, Fons Leroy, Rik Van Bruggen, Rene Steenvoorden, Philippe Gosseye, Philippe Rogge, Kosta Peric, Penni Geller, Paul Daugherty, Michael Kogeler, Vijay Gurbaxani, Luc De Vos, Lieven Haesaert, Johan De Geyter, Inge Geerdens, Jef Staes, Geert Noels, Peter Claes, Chris Van Doorslaer, Abdella Bouharrak, Bill Chang, Amanda Jobbins, Luc Bleyaert, and Adam Pisoni.

Special thanks go to my longtime business partners, Luc Osselaer and Fonny Schenck.

www.peterhinssen.com

PREFACE

WHY NETWORKS MATTER

Since I was a kid, I've been obsessed with maps. I adore the layout of old maps depicting mountains and oceans, cities and roads. I find it fascinating how world history can be seen through the shifting borders of maps over time. But nowadays, what amazes me most is how old cartographers got it wrong.

One of my favorite examples is the way California has been depicted, having lived there for part of my childhood. For the longest time, travelers believed that California was an island, separate from the American mainland.

My favorite type of map is the road map, whether it shows layouts of city streets or depicts how railroads and highways, roads and turnpikes, lanes and pathways curve through the landscape of countries and continents. For me, the best part of the *Lord of the Rings* books is the amazingly intricate maps drawn by J. R. R. Tolkien, which let you follow the trail of the merry fellowship on their journey to Mordor.

But moving on from my hobbit habits, when you look more closely at the topology of maps, it's clear that roads form a network. The infrastructure of how cities operate, their capacity and connections, tell the story of the economy and influence of a region. You can see which cities are at the heart of an economy and which are at the edge. You can understand how influential a town is, what its main economic arteries and roads are, and how it influences, stimulates, and feeds neighboring towns and hamlets.

I now spend countless hours poring over Google Maps and Google Earth. I love exploring regions that I've visited or will visit, ogling the enormous beauty and complexity of the roads and canals, and the intricacy of the routes and freeways. All of this was made by man.

When it comes to the mountains and lakes, the volcanoes and harbors, the plateaus and cliffs, we had nothing to do with those. But all the rest, that was us. We created these networks that carry our cars, wagons, trucks, boats, buses, bicycles, and pedestrians. There was no grand plan, as is made obvious by the myriad twists and turns and the fractallike complexity of the network.

There are exceptions, of course. The German autobahns were created through a grand scheme, and the U.S. interstate highway system[1] copied the idea. These pathways seem more logical than the rest of the infrastructure, which seems almost organic. But their impact is immense. The U.S. interstate system helped shape the United States into a world economic superpower and a highly industrialized nation.

In newer economies like that of the United States, you tend to see more deliberate structure and order. Cities such as New York or Chicago are laid out on grid patterns, for example. It's much easier to find the corner of 57th Street and 5th Avenue in New York than the intersection of Avenida de Lanzarote and Calle Nicaragua in Madrid.

Over hundreds, sometimes thousands, of years, these networks have grown and connected cities and towns, and have become the lifeblood of our economies. We hardly realize that anymore. On the contrary, we notice it only when it fails—when we're stuck in traffic, trying to get into or out of a city, and cursing the designers of these blasted roads, traffic signals, or bypasses.

But the truth is, these infrastructure networks are the heart of our economies. They transport employees to their jobs, children to their schools, produce to the warehouses, and eBay purchases to your front door. Without these networks, there would be no economy, no global society, no wealth, and no prosperity.

But today, the roads on a map tell only a tiny part of the story. They tell the tale of the physical part of our economies—as Nicholas Negroponte[2] would say, the "transport of atoms."

Unseen by the naked eye, superimposed on these visible networks are countless invisible networks: wireless networks that connect people, phones, tablets, and other devices; radio and television networks that relay music and video across the globe; telecommunications networks that connect satellites, bringing continents and remote areas together; electricity networks that bring power from wind turbines at sea to homes, or watts from nuclear plants to factories; information networks that relay more and more content and knowledge at increasingly breakneck speeds.

What we see when we gaze upon our planet are manmade infrastructure networks that took hundreds of years to form. And if you could observe the spectrum of communications networks on top of that, you would find layer upon layer of networks, created over the last century, that have grown to a similar level of complexity and delicacy. All are expressions of humankind, our collective actions as a civilization.

I have become obsessed with networks: their absolute power and their pure simplicity.

So this book is about networks: why I have come to believe in them as the most fundamental driver of progress, and why we have to understand networks if we want to survive the next era of society.

I believe that networks will win. Networks always win.

We're witnessing a revolution in the way society works. Many people, including me, thought at first that this was because of digital technology— that we were witnessing the advent of a new society, with digital natives leading the way and the toddler who can swipe an iPad as its poster boy. We were wrong.

It's not because of digital technology. It's because of networks.

What is happening in front of our eyes is that everything is becoming connected to everything else. Information is flowing through networks with greater intensity, and that completely changes everything. Markets are disappearing, becoming networks of information with the customer at their heart. And if the outside world becomes a network, companies will have to follow suit.

That's the punch line. Plain and simple.

If you understand networks, you will understand the future.

Peter Hinssen

LET'S GET STARTED

ABOUT *THE NETWORK ALWAYS WINS*

In this book, we'll explore what it means to become a network organization, and how to influence consumers who behave as active elements of a network, not passive targets of a market. We'll examine what it takes to match the speed of network thinking, and how to transform your organization to maintain its relevance in the age of networks.

But on a deeper level, this book is about much more.

This is a book about **speed**—about what happens when markets move faster than you can control, or even observe, so that you will have to move faster without either absolute visibility or absolute control.

This is a book about **consumers**—about the shift that takes place when they stop simply consuming and start interacting, and when markets stop being markets and become networks of intelligence.

But this is also a book about **organizations**—about what will happen when we have to adapt the inside of our companies to markets that are changing faster than ever before. We will have to adopt this network thinking inside our own organizations.

This is not a book that is meant to frighten you. It was written to make you more alert—to open your eyes to what is probably the biggest shift we've ever seen in society, the biggest shift ever to hit the world of commerce, and the biggest shift in how we build organizations.

This is really a survival guide designed to teach you the new rules of the land: about the new patterns, the new cultures, and the new behaviors, and how to reinvent your company after your market has flipped and turned network.

This is not a technical book. Don't expect any detailed discussions of the technical aspects of networks, the routing of TCP/IP packets, or the tweaking of 4G wireless routers. This is a business book.

We'll dive into some physics and mathematics here and there—first, because the stories are really good, and second, because they will make you feel pretty good about yourself for having read a book with some equations in it.

But above all, this is a book about our world and what is happening all around us, a book on how we—and our children—can survive in a world in which information is ubiquitous. We will have to figure out how to dismantle the old industrial structures that now form the pillars of our economy, and rethink how companies, organizations, and society function. And perhaps the answer to understanding our future is right in front of our very eyes. We only have to gaze upon ourselves and observe the networks that we have built.

WHAT I GOT WRONG IN *THE NEW NORMAL*

I've become obsessed with S curves. A typical S curve has two really interesting parts. At first, the movement is slow, but then it suddenly picks up and starts to rise exponentially. Think about the uptick in smartphone owners recently, or the number of users on Facebook. The speed increases and grows exponentially. There's no stopping the momentum, and it promises to go on forever.

But then the magical moment happens, and a flip occurs. The S curve starts to slow down. Growth is still fast, but it's moving in the opposite direction from the first half, and when we get into the second half, we see a stabilizing factor where growth eventually tops off.

Many markets have followed this phenomenon. One of the best studies of S curves was done by Carlota Perez, technology expert and author. She applied these curves to the growth in railways, the automobile market, the Industrial Revolution, the rise of electricity, and the growth in computing power.[1]

That's what caught my attention. I wrote *The New Normal* on the rise of digital and what happens when digital passes the halfway mark and becomes normality. In other words, this was digital's path along the S curve.[2]

In the beginning, when digital was still new, its growth was slow. Only the true nerds were hunched over primitive microcomputers like the Apple II, the Commodore 64, and the Atari. Normal folks would look at them with a mixture of pity and contempt.

But then the market caught fire. The pace picked up, and suddenly everyone had a computer: first desktops, then laptops, and now tablets. And digital became normal. Older people still remember when digital was a meaningful adjective. Younger people don't. Everything is digital. Analog is the exception.

And then we get into the second part of the S curve, a world where we take digital for granted. Companies that don't adapt become obsolete in a world that is governed by Google and Twitter, and where the breakneck speed of keeping up with digital is rapidly revealing the incredible slowness and immobility of some organizations.

The New Normal was a book about what happens when digital becomes normal—about the flip to a digital world.

But it also got me hooked on S curves. And I now believe that this is much bigger than just digital. This is way beyond the world of iPhones, Spotify, and Facebook. Technology is just one example of markets that *flip*. Digital was just the appetizer.

The flip is a broad concept about what happens when markets start to move faster than companies.

In one sentence: this book is about how to survive the flip.

CHAPTER 1

———

THE AGE OF UNCERTAINTY

Building a culture of experimentation
means implementing an attitude toward risk
that will be necessary if we are to survive in a world
in which strategy becomes fluid.

———

We would love to build a model that describes everything about our markets, customers, and organizations, and use that model to build the perfect strategy for the future. Unfortunately, that doesn't work, and it never will. The future is characterized by VUCA: volatility, uncertainty, complexity, and ambiguity. Therefore, no model will ever suffice. Instead, we have to focus on building agility, acting fast, and nurturing creative thinking by stimulating our companies' networks and adopting a culture of experimentation.

The Holy Grail for many companies is to build the "ultimate knowledge tool." This is a mythical spreadsheet that companies hope to use in order to build an all-encompassing model of their enterprise. They could feed into it all the inputs of their business, insights into the behavior of their customers, and all the relevant knowledge about trends in their markets. From this ultimate shiny model, they would be able to deduce the perfect strategy.

Nobody has yet built such a spreadsheet. No one has found the Holy Grail, either, but that doesn't mean that people aren't still looking for it after two thousand years. In the same way, companies seem to be constantly in search of the perfect strategy.

In fact, strategy is the number one category in the extremely lucrative management book market. Innovation is number two, because people quickly realize that they have to come up with something that is truly unique if they are to survive. Third on the bestseller list is self-help guides, written to help the disenchanted crowd of depressed businesspeople who have learned that there is no ultimate strategy and no magic recipe for innovation.

THE THEORY OF EVERYTHING

We humans seem to be obsessed with modeling. It's our way of figuring out how things really work, and our attempt to do that by using formulas, concepts, and mechanisms. The history of physics is thousands of years of trying to figure out the elusive "theory of everything" to explain it all.

Physics is still trying to figure it out, and every time the physicists think they have it nailed, they find that they have to dig deeper and build more models, like the Russian matryoshka nesting dolls. Newtonian physics worked like a charm, explaining everything from the movement of planets to the falling of apples. It seemed to be pretty much perfect—that is, until it was applied to things that were very small. For that, we needed quantum mechanics. Or how about for things that go very fast, such as light? For that, we invented the theory of relativity. Every time we dig a little deeper into physics, we understand a bit more about how the universe actually works, and we adapt our "model" accordingly.

Physicians have been trying to understand our bodies and the complex workings of our brains for centuries. And it's true that the medical profession now knows a considerable amount more than it did in the days of bloodletting and lobotomies. But we are still light-years away from being able to model the human brain. Yet we humans are a persistent bunch. We won't stop until we've built that ultimate model of our gray matter.

In business, we also try to model everything. For years, economists have been trying to model markets, companies, processes … anything, really. They've been trying to create models that would explain market behaviors and models that would even explain the weather. They've even tried to create a "magnum model" of models.

THE RAND CORPORATION

One company that has always fascinated me is the RAND Corporation. This secretive organization, based in Santa Monica, California, was formed in the wake of World War II by the U.S. Department of Defense, the Air Force in particular. The Air Force viewed itself as being much more sophisticated than the Army, from which it had been spun off. The Army had troops and tanks, while the Air Force had all kinds of cool stuff, from planes to nuclear technology. In order to maintain that edge—and to maintain control of nuclear strategy—the Air Force needed a research arm. That's why the RAND Corporation was created.

The RAND Corporation (short for research and development) was set up in 1946. Its primary goal was to help the U.S. Air Force define a strategy for the long-term planning of future weapons. In other words, during the cold war, it would plan for nuclear war.[1]

Like the Manhattan Project before it, RAND recruited the smartest people on the planet, including more than 30 Nobel Prize winners. Since its inception, RAND has drawn upon top scientists, mathematicians, and physicists who believed in the ultimate model—and in the ultimate chance to use mathematics to shape the universe and the future.

The charged atmosphere at the RAND Corporation during the fifties must have been exhilarating and intoxicating, with the sharpest minds of the "free world" working to figure out how to remain free—and win a nuclear war.

One of the most colorful and influential figures in the RAND Corporation was a man named Herman Kahn. Stanley Kubrick immortalized him as "Dr. Strangelove" in 1964.

Herman Kahn was born in New Jersey, the son of Jewish immigrants from Eastern Europe. He was raised in Los Angeles and attended UCLA, where he got a degree in physics. He was working on a PhD at Caltech, but he had to drop out for financial reasons (not for lack of talent). Rumors were that he had the highest IQ ever measured. Kahn became friends with Samuel Cohen, the inventor of the neutron bomb, who introduced him to the RAND Corporation. He was recruited soon afterward, in 1947. Kahn took to RAND like a duck to water.

He quickly became a major player at RAND, with a unique style. A master showman, he would deliver 12-hour lectures spread over two days, using huge sets of charts and slides.

For more than 10 years, Kahn developed a novel technique known as systems analysis. The idea was to build the most perfect model of a strategic question by first analyzing and then synthesizing a model that could be used to run different scenarios. The verbs *analyze* and *synthesize* have Greek origins, where they mean, respectively, "to take apart" and "to put together." That's exactly what Kahn wanted to do for nuclear strategy. The result would be a model of the cold war that could define not only what the endgame with the Russians would be, but also how much it would cost. [2]

To Kahn, it was like playing a giant game of Risk. But given the mighty computers that he had at his disposal and the enormous impact that his models had on real nuclear strategy, this game was frightfully real. By putting into his models the quantities of bombers and radar systems, and the location of airfields and missiles, Kahn could calculate how many civilian deaths would result from various nuclear scenarios. He could determine how many cities would be wiped out and how much collateral damage there would be. Kahn's model of the cold war directly influenced how many billions of dollars were allocated to the U.S. Department of Defense. But he was also calculating just how many hundreds of millions of people would perish in the event of a full-blown nuclear war.

He finally published his magnum opus, called *On Thermonuclear War*, in 1960, the title being a reference to the classic nineteenth-century work by Carl von Clausewitz, *On War*. His boss thought Kahn had gone too far, and told him that the work "should be burned." However, the book provided amazing insights into the model that the RAND Corporation had built. In the first three months, *On Thermonuclear War* sold more than 14,000 copies, and it was read with great interest on both sides of the Iron Curtain. The Soviet newspaper *Pravda* trashed the book, calling RAND the "academy of science and death." However, U.S. Secretary of State Henry Kissinger was a great admirer of the book, and of Kahn.[3]

The amazing thing about the model that Kahn built was that it described a post-nuclear war world—a world in which, depending on the variables, hundreds of millions had died or "merely" a few major cities had been destroyed. Kahn argued that life would go on, and that the "survivors would not envy the dead." Just as Europe had managed to continue after the devastatingly deadly effects of the Black Death in the fourteenth century, Kahn believed that people would cope, and that humanity would prevail.

Understandably, Kahn's theory was highly controversial. Not only did he use his model to think about the world after a nuclear war, but he was also incredibly provocative in his solutions for coping in a post-nuclear holocaust era. For example, he suggested grading food based on the amount of radioactive contamination it contained. The most highly radioactive food should be consumed primarily by the elderly, who would probably die before the delayed onset of cancers caused by radioactivity. Kahn was a nuclear provocateur. But he was a scientist first, and in the fifties it was very plausible that there would be a nuclear arm-wrestling match between the United States and Russia.

As Kahn put it, he was "trying to design a system capable of meeting contingencies which will arise five to fifteen years into the future." To put it differently, the RAND Corporation was trying to predict how the world would evolve—exactly what every company that is chasing that Holy Grail of the "Ultimate Excel" is trying to do.

THE ULTIMATE MODEL

Kahn took advantage of the rise of digital computers, which were building up enough horsepower to be able to do those calculations, to run all those simulations, and, ultimately, to take the human side out of the equation. Certainly, in the context of predicting the outcome of nuclear war, with possibly hundreds of millions of civilians dying, the human factor in decision making was often disturbing. As Kahn put it, "under these circumstances, competent, honest people often don't do very well."

And here we come to the heart of the matter. We seem to be set on building the ultimate model—one that will guide us to the ultimate strategy and eliminate the human element in decision making.

Kahn believed that if computers got better, if technology got stronger, then he could build better and better models and get closer to that Holy Grail of the ultimate model. Indeed, during this period, the capabilities of computer systems were advancing by leaps and bounds. Kahn observed an "exponential increase in the state of the art of [computer] technology in the past twenty years." Little did he know that this was only the beginning of the exponential curve of technology dictated by Moore's Law.[4]

Kahn also understood that the technologists who were working on these new computer systems had a habit of overestimating their ability to apply new ideas in the immediate future, and underestimating the impact of their technology on the long term.

STEPPING INTO THE UNKNOWN

After a falling-out with the RAND Corporation in 1960, Kahn left and founded the Hudson Institute, then located in Croton-on-Hudson, New York. In his words, the Hudson Institute was to be a "high-class RAND." In 1967, Kahn produced a fascinating report from the Hudson Institute, called *The Year 2000: A Framework for Speculation on the Next Thirty-Three Years*. In it, he described 100 technical innovations that were very likely to happen. Among them were:

74. Pervasive business of computers
81. Personal pagers (perhaps even pocket phones)
84. Home computers to run households and communicate with the outside world[5]

If you thought that Bill Gates or Steve Jobs invented home computers, or that it took until the World Wide Web to envision something like the Internet, think again. Kahn had amazing insight, even predicting, as item 67, the "commercial extraction of oil from shale." He was a bona fide genius, disliked by many for the controversies he created, but a true pioneer of models. He passed away in 1983, at the age of 61.

But Herman Kahn also saw the limitations of models—that they could get you only so far. He was a great student of the history of war, and he was fascinated by the mistakes the French made in dealing with Germany in World War II. In his words: "The French had more and better tanks than the Germans, about as many planes, and given their fortification systems, seemingly as good an army. The Germans were just better in the imponderables." An *imponderable*, as the dictionary defines it, is "that which is difficult or impossible to assess"—in other words, the unknown. As Kahn describes it, by conventional measures, the Germans would have had no chance against the French at the advent of World War II, but the Germans avoided a direct assault on the famous Maginot Line of French defense by violating the neutrality of Belgium and penetrating France from that direction on May 10, 1940. By June 22, they had invaded most of France, and the French surrendered. The invasion was not planned by any model, but by making clever use of an imponderable. Certainly the French never thought of that possibility.

Kahn realized that in the end, models are only models, and you can't predict every imponderable, no matter how much computing power you have. He actually told the U.S. Department of Defense that "the most important thing to do is to see that we maintain a great deal of flexibility in our military capability, and have available many options in order to meet a variety of contingencies." In other words: prepare for the unknown by keeping as many options open as possible. So much for those powerful models.

If you look at military strategy after the cold war, which was still being devised by the great military think tanks like the RAND Corporation, the planners started to take these imponderables very seriously. Actually, they now pretty much depend completely on understanding the world of the imponderables.

Today, the concept of imponderables is at the heart of military thinking. Strategists realize that the world can change instantly. The end of the cold war came like a bolt out of the blue. And no Kahn model could have ever predicted the events of 9/11. As a result, current military strategy is now based on the concept of VUCA.[6]

VUCA

Let's explore what the acronym VUCA means to corporations.

V stands for volatility. The word *volatility* comes from the Latin verb *volare*, which means "to fly." It basically means that things tend to vary often and widely. In today's world, things are changing faster and faster, and stability often seems unattainable. After the financial crisis and the economic turmoil of the past few years, many people have been wondering, "When will things get back to being stable again?" The answer could be never. Perhaps we're living in times of greater and greater instability, more and more turbulence, and increasing volatility. Stability is flying away.

U stands for uncertainty. Many of us grew up trying to build a world based on security—things that we could count on, bank on, and build on. Now, more and more, we seem to be surrounded by uncertainties. We're uncertain about where the economy is heading. We're uncertain about where technology is heading. We're uncertain about whether our companies will survive, how our customers will react, and how our markets will evolve. We seem to be heading toward a world in which we have more uncertainties than certainties to deal with.

C stands for complexity. The world has become incredibly complex, and little things can have a huge effect. Scientists have been studying this dynamic, in a field that is often labeled *chaos theory*, for years. It's often characterized by the notion of the *butterfly effect*, a term coined by meteorologist Edward Lorenz back in 1963. Lorenz was trying to predict the weather by building complex mathematical models of weather systems. (You could say that he was the Herman Kahn of weather forecasting.) He observed that although the computers that ran his models were getting better and faster, and his models were getting better and more complex, his ability to predict the weather wasn't improving much. He observed chaotic behavior in his weather systems, in which tiny differences in a system could trigger vast and unexpected results. In a 1972 paper, he described this by saying that a flap of a "butterfly's wings in Brazil could set off a tornado in Texas." Today we see more and more of these chaos-theory effects in a world of ever-increasing complexity.

A stands for ambiguity. Ambiguity means that you can interpret things in more than one way—that things depend on their context, and that you have to interpret the whole picture in order to understand something.

Ambiguous is not the same as vague. It means that things are not always black or white, one or zero. Too bad for all the digital zealots, but the world can't always be boxed in. Ambiguity means that questions don't always have a single "straight answer."

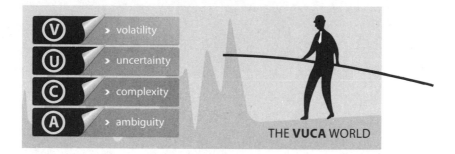

> volatility
> uncertainty
> complexity
> ambiguity

THE **VUCA** WORLD

But what does VUCA mean for corporations? What does it mean for companies with armies of workers laboring in structures built on the notion of long-term thinking? Or for those who have built strategies on the basis of consumer models, market models, and growth models based in Kahn-like reasoning?

The result is that in the VUCA world of today, business strategies have become more fluid. The traditional five-year plan makes no sense to companies whose world is changing faster and faster. What use is a five-year strategy for the newspaper industry, which is watching its markets disappear at the hands of Twitter and Craigslist? What use is a five-year strategy for the television industry, which is seeing its markets flip completely because of YouTube, Netflix, and Hulu?

The reality is that companies now update their strategies more frequently than ever before. They have to, if they want to survive. I heard one CEO tell me, "We still have a five-year strategy. Actually I have a new one every three months."

This is the era of "fluid strategy," and it means that companies have to act more quickly—and be more agile—than ever before.

VACINE

As a matter of fact, as an antidote to VUCA, I've coined another acronym, VACINE. It's designed to inoculate organizations against this mad VUCA world in which our old models have grown increasingly useless.

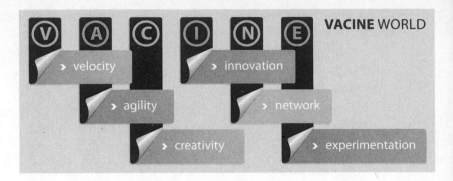

V is for velocity. It's clear that companies today have to move quickly. They have to be able to get their act together and move at the speed of the market, their customers, and the ecosystem in which they operate. The next chapter of this book is completely dedicated to speed, and how fast companies have to be able to move nowadays.

A is for agility. Speed alone is not enough. Agility is about being able to move quickly *and* swiftly—to be able to turn on a dime. It is about the capacity to change, to respond, and to transform. Agility is all about being nimble and being quick on your feet. In the sport of boxing, there is the phrase "roll with the punches." For companies, this means coping with and withstanding adversity by being flexible.

One of my favorite examples of agility comes from the French telco company Free. Its biggest competitor, Bouygues Telecom, launched a big promotion campaign. The latter was probably convinced that it would be able to steal away quite a few of Free's customers. But, powered by big data intelligence and a firm knowledge of how low the company was able to go (pricewise, of course), Free launched a cheaper counteroffer just one hour later. Talk about speed of reaction. And Free got quite a lot of visibility from this, stating that it was not the cheapest in the market for merely one hour.

Some of the largest and most innovative companies I know started out as something completely different. But they were smart enough to recognize a better opportunity and agile enough to make the flip. IBM (founded in 1911) originally sold commercial scales and punch-card tabulators; later on, it sold massive mainframe computers; and today it is all about software and consulting and IT services. Corning (founded in 1851) used to create the glass enclosures around Thomas Edison's lightbulbs. Now it specializes in optical fibers, cell phone covers, telescopic mirrors, and television screens.[7] The William Wrigley Jr. Company started out as a soap and baking powder company in 1891. In 1892, it offered chewing gum with each baking powder package. When it saw that the chewing gum was growing more popular than the baking powder, Wrigley reoriented itself. There can be a lot of luck and serendipity involved in innovation, but the real trick is being able to turn with the opportunities that present themselves. A lot of larger companies are too bureaucratic and too slow to manage that.[8]

C is for creativity. In today's whirlwind economy, creativity has become an asset to cherish. Companies that fail to maintain a creative edge fall fast and hard. If you examine the stumbles of once-giants like Nokia or RIM BlackBerry, you see that they tumbled soon after they suddenly lost their creative flair. Legendary for adding his own creative touches to Apple products, Steve Jobs talked about "connecting the dots." As he phrased it, "You can't connect the dots looking forward; you can only connect them looking backwards. So you have to trust that the dots will somehow connect in your future. Because believing that the dots will connect down the road will give you the confidence to follow your heart even when it leads you off the well-worn path. And that will make all the difference."

I is for innovation. Innovation has become the lifeblood of organizations. "Innovate or die" was once the mantra of the high-tech industry, but today the rate of change in many markets is picking up so fast that the motto has become mainstream. Innovation, the ongoing effort to develop new ways to solve current and future problems, is the main driver of differentiation and sustainability. It determines the ability of an organization to remain relevant amidst continuous and constant change.

N is for network. I believe that a major shift will occur in the way we adopt the network within our organizations. This book focuses on the rise of network effects in markets and what this means for companies. In a nutshell, I believe that markets are turning into networks of information, with the customer at the heart of those networks. This means that we can't "market" anymore; instead, we have to be able to "influence" networks of information. And if the outside world becomes a network, then the inside has to become a network as well. Companies will survive in the age of networks only if they become networks of innovation internally.

E is for experimentation. Finally, and perhaps most important, companies will have to learn to adopt a culture of experimentation. In many companies, failure is an absolute taboo. Companies today are set up to execute strategy and implement five-year plans, but they're still learning the art of experimentation. Start-ups, on the other hand, already experiment all the time. They are not afraid to fail.

Some of the most successful start-ups in Silicon Valley have adopted the "fail fast forward" principle, under which they encourage people to try many things, ruthlessly kill what doesn't work, and amplify what has potential. The purpose of failing fast is to understand what doesn't work and to learn from your mistakes. Building a culture of experimentation means implementing an attitude toward risk that will be necessary to survive in a world in which strategy becomes fluid. Some of the most avid advocates of this approach can be found at FailCon, a one-day conference for technology entrepreneurs, investors, developers, and designers who study their own and others' failures in order to learn from them.[9] You have to be a certain kind of brave to be prepared to admit in front of your peers that you made a mistake. But just imagine the energy and the insights that can arise from such a "learn from your failures" network.

But it is not only start-ups that are embracing this approach. An increasing number of larger or growing companies are using the same kind of "fail fast" methodology. Susan Wojcicki, Google's senior vice president of advertising, understood this very well. She incorporated "never fail to fail" into her "eight pillars of innovation" a while ago. Spotify, for instance, performs small-scale "quick and dirty" experiments that are terminated really fast when they don't work.[10]

Pixar too, one of the most creative and innovative companies in the film industry, has a unique take upon failure. In the words of Pixar cofounder Ed Catmull: "[Many people] think it means accept failure with dignity and move on. The better, more subtle interpretation is that failure is a manifestation of learning and exploration. If you aren't experiencing failure, then you are making a far worse mistake: you are being driven by the desire to avoid it. And, for leaders especially, this strategy—trying to avoid failure by outthinking it—dooms you to fail."[11]

The rest of this book will help you put VUCA in perspective, and will hopefully help you implement VACINE in order to cope with this ever-faster pace of change in our markets and our companies.

The days of long-term strategy are probably over, never to return. It's about being nimble and fast, responsive and agile. As the great philosopher Mike Tyson once said, "Everybody has a plan until they get punched in the face."

As Herman Kahn wrote, after describing the horrible outcomes of global thermonuclear war, "We sincerely hope that the reader has a slightly uncomfortable feeling at this moment." It was a classic example of Kahn's black humor. I can imagine that you also feel slightly uncomfortable after reading this chapter, its descriptions of VUCA, and the end of long-term planning. But rest assured, the aim of this book is not to alarm you, but to guide you through this Age of Uncertainty.

CHAPTER 2

SPEED, AND WHY THE THEORY OF RELATIVITY MATTERS

We find ourselves citizens of the *era of now*. Trends are a thing of the past. Real-time insight into what's happening is good. Foresight into what we may expect is even better.

Products, markets, and behaviors are evolving faster than ever. Is your organization's ability to change keeping up? Is your internal clock ticking fast enough?

A BRIEF HISTORY OF TIME

Have you ever noticed that things are going faster than they used to? Have you ever had the feeling that the world is speeding up? That things just keep moving more and more quickly?

You're not alone. Things are happening faster. Trends materialize faster. Novelties wear off sooner. News gets old faster. Speed is the name of the game, and our whole world seems caught up in an ever-faster pace.

In business terms, this is not necessarily a problem as long as you can move faster than the market. One of my favorite quotes is by Formula One legend Mario Andretti, who summed it up perfectly: "If everything seems under control, you're just not going fast enough."

SPEED

One of the most ridiculous film plots of all time is from the 1994 movie *Speed*. Imagine being a scriptwriter and pitching your idea to a major studio executive: "Well, there's this bus, see, and you have to keep driving it above 50 miles per hour or it explodes." It's a small miracle that the movie was made at all. The bigger miracle was that it was so successful. Anyway, despite the dubious plot, people seemed to love the film.

But the interesting question that this film triggers is one that many companies should be asking themselves today: "In the ever-changing and fast-paced world that we live in, how fast do we have to drive the 'bus' (our company) so that it won't explode?"

FASTER, FASTER, FASTER

The rate of change is accelerating. The question is whether this is just perception or reality.

We've all heard older folks talk about how time goes by faster as they age. We've probably all heard our grandparents say things like, "Oh, my, is it July already? Now where did the time go?"

It could be that the older we get, the more intense our perception of time passing by gets. Or perhaps we become painfully aware that we don't have an infinite amount of time left.

But it's not just a personal appreciation of fleeting time that we're experiencing today. We seem to be caught in a whirlwind of speed on a global economic scale. Take the wonderful world of telephones. I remember vividly the importance that a telephone had in our household when I was a child. The home phone—mind you, a single phone for the entire family—was a prominent feature of our home, firmly connected to the wall. We had that same phone for ages.

Enter the world of the mobile phone. I've probably gone through more mobile phones than pairs of shoes in the last five years. Mobile operators seem to be churning out new subscription packages on a monthly basis. It's getting so bad that my children are terrified of having to face their classmates with a smartphone that is clearly last year's model. The horror!

The velocity of the mobile industry also caused the rapid rise and even faster decline of players such as Nokia and Research in Motion (BlackBerry). The phrase "Here today, gone tomorrow" has been replaced by "Here today, gone today." Nothing lasts forever, but many things don't even survive the frenzy of today.

THE ERA OF NOW

The result is that not only are products and markets moving faster and faster, but our behavior patterns are shifting as well. Everything has to happen *now*. We talk about instant feedback, instant messaging, and instant gratification.

Time pressure is all around us. We can't miss new information or we'll fall behind. We can't miss an appointment, a conference call, or a meeting. We have become slaves to a clock that is ticking faster and faster. We find ourselves citizens of the *era of now*. Trends are a thing of the past. Real-time insight into what's happening is good. Foresight into what we may expect is even better.

We seem to be less and less inclined to "take the time." Take the example of our information consumption. We are constantly engulfed in information, with the volume and velocity of information only increasing. Years ago, I stopped reading a printed newspaper and shifted to reading news online. I soon noticed that because of the speed and quantity of information, I was just scanning the headlines instead of reading entire articles. I find it rarer and rarer that I take the time to read an entire book, or even an entire magazine. (Kudos for making it so far into this book!)

According to Nicholas Carr, the result of information overload is that we distort our brains so that we start to consume smaller amounts of information (hence the success of text messaging and Twitter). Our brains, according to Carr, are being rewired so that we are able to absorb more information, but only shallowly.

We moved from solid meals to fast food. Now we are down to snacks.[1]

CLOCK SPEEDS

At the end of the 1990s, Charles Fine investigated the rate of change of companies and markets. Back then, computer shoppers would seriously evaluate the clock speeds of different processors. Today, most people don't care how fast the internal clock ticks inside their devices anymore. And they shouldn't.[2]

Fine raised an interesting question: "What would the internal clock speed of a company be?" In other words, how fast does a company tick? How fast does it cope with change? Is it possible to compare different companies based on their clock speeds?

Mind you, Charles Fine asked this question in the days before the Internet. But because companies are now all living in a connected world, all part of the network of information, it's now much easier to compare the reaction speeds of various organizations. We can therefore take Fine's ideas a little further.

Consider the market that you're in. What's its clock speed? That is, at what rate does your market change?

Some markets are clearly moving faster than others. Consider the world of mobile technology, where we see breakneck clock speeds. Or the world of music, where YouTube and iTunes have transformed not only the nature of the music business, but also the speed at which bands are discovered and songs can go viral. (Going viral in a networked society means that you can propagate faster than the average speed in the network—basically, that you're speeding).

But there are some markets that are still relatively slow. I've had the pleasure of spending time with construction companies that are building highways or high-speed train lines, or dredging harbors in the Middle East. Those markets move much more slowly than the mobile and media industries.

But whatever the speed of your market, consider how that speed is evolving. Is your market moving faster than it did a decade ago, and do you expect it to be moving even faster in the next decade? If so, you're dealing with an accelerating external clock.

INTERNAL CLOCK SPEED

Now let's return to the clock on the inside of your company. One of the first questions I ask when I visit a company is: "How long does it take to get anything done around here?" This is not meant as a loaded question, but as a serious one. Inside your organization, how quickly does an idea evolve from a brainstorm into a marketable service or product?

Most likely, your company's internal clock has accelerated over the years. If you could look back to the time when you first joined your company, you'd be amazed at how slowly things moved then. Think about how fast your portfolio of offerings has evolved, and how quickly you've introduced innovations to the market.

internal clock   external clock

Internal clock speed itself is irrelevant nowadays. All that matters is whether you're moving faster than your market: Is your internal clock outpacing the external clock?

In the past, companies had enough time to innovate, adapt, and evolve. As a matter of fact, they imposed the rate of change. Automobile manufacturers determined which novelties would be in our cars; telephone companies decided when new service features would be available; soda companies dictated which flavors you could buy.

Those days are long gone. Markets have a heartbeat of their own. Today, market dynamics and market speed are accelerating. The outside world has begun to pulse at an ever-increasing pace, fueled by the power of the network. But many companies are falling behind. In fact, most are.

THE CLOCK SPEED CONUNDRUM

If your company is moving significantly more slowly than the market, that's a *major* problem—a problem that, if left unfixed, will quickly become a "we're all going to die" kind of problem. The world of business is filled with examples of once-successful companies that have fallen behind and didn't quite make it.

One of the best examples in recent history is the complete reshaping of the music business with the introduction of the MP3. Soon after that came iTunes, taking a wrecking ball to the business model of record companies. Understand that this was an industry that was used to change. Before the MP3, there had been the CD. And before the CD, there had been vinyl. And cassettes. And the eight-track. There had been plenty of minor revolutions. So the music industry wasn't afraid of technological innovation. It had technical people to deal with that.

But the rise of the Internet, and the ability to download music effectively and securely, introduced a kind of velocity to the business that it had never seen before. All of a sudden, a new delivery mechanism (MP3 over the Internet) meant that totally new business models were opening up at a rapid pace. Steve Jobs—the incredibly talented observer of paradigm shifts—smelled that the music industry would not be able to move fast enough. He quickly established iTunes as the dominant player before the music industry could move beyond its internal quarrels.

The players in the music industry had a clock speed that was fine for the old days. iTunes moved in with a clock speed that was in tune with the external clock speed of the Internet. Traditional music industry players weren't just defeated; they were made irrelevant.

To survive in the era of networks, you must understand the rate of change outside your organization as well as the capacity for speed within it.

THE THEORY OF RELATIVITY

Something similar happened in physics more than a century ago. It took the genius of Albert Einstein to write down the special theory of relativity in 1905, which led to the famous $E = mc^2$ equation. The brilliance was his postulation that the speed of light is an absolute constant. This, of course, led to countless science fiction writers imagining the possibility of moving faster than light. There would be no Millennium Falcon allowing Han Solo to jump to hyperspace if Einstein hadn't produced his famous paper.

Years later, in 1916, Albert Einstein introduced his now famous special and general theory of relativity. This amazing period in physics at the beginning of the twentieth century was a whirlwind of scientific research and discovery centered around what could happen if we started to explore the world at the extremes—the world of the very fast, and the world of the very tiny.

Before that, the laws of physics had been described by Newton. He had accurately described the world around us: how it functioned, how the elements of it interacted with one another, and how apples fall on our heads. His laws of physics were the basis of the Industrial Revolution, which gave us steam engines, trains, machines, and mass manufacturing.

But something really strange happened when you tried to apply those laws to things outside of our normal observations—to things that go really fast, at the speed of light. That's exactly what Einstein wanted to figure out: what would the world look like if you could ride on a beam of light? He predicted that it wouldn't look at all like the world we know.

And he was right. It turns out that Newton's laws simply don't function when objects move really fast. It also turns out that Newton's laws stop working when we look at the really tiny world of subatomic particles. All of this informed our understanding of the world of quantum mechanics. But that's a different story for a different book.

The point here is that when we change our perception beyond the familiar world we know, our fundamental principles of operation are bound to change.

Therefore, as we evolve into a world in which markets move faster than organizations, we're bound to have to rethink what is normal. It may very well be that our normal mechanisms and rules don't work anymore. Perhaps we need to reinvent how organizations function when we cross the boundary where external clock speeds outrun internal clock speeds.

BREAKING THE SOUND BARRIER

But we don't even have to go that far, or that fast.

Toward the end of World War II, there was a frenzy as we tried to break the vaunted sound barrier. Was it possible to fly a plane faster than Mach 1—the speed of sound?[3] It seems like a silly question now, but back then, the question was both real and urgent.

Who was going to be the first to break the sound barrier? The British, Germans, and Russians all wanted to be the first. That didn't happen, although they tried very hard and their test pilots often paid with their lives. But where everyone else failed, the Americans succeeded. In 1947, U.S. Air Force pilot Chuck Yeager was the first person to fly faster than the speed of sound. He did it in his experimental craft, the Bell X-1, christened the Glamorous Glennis after his wife. Captain Yeager was one of the best test pilots in the world, and it took superb flying skills to control these flying rockets. But even after Captain Yeager's achievement, it was decades before supersonic flight was really understood and mastered. To this day, we're

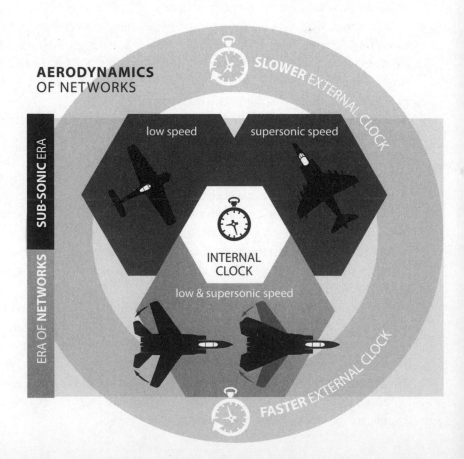

still waiting for a commercial supersonic success story. But in the military domain, supersonic is an absolute winner.

For the purpose of this book, the really interesting aspect of supersonic airplanes is how they are built. They require a completely different way of thinking about design. Supersonic planes behave differently, requiring different wings, different design principles, and different mathematics and physics. With today's knowledge of aerodynamics, it's pretty straightforward to design a plane to fly slower than the speed of sound, and even to design a plane that can fly faster than the speed of sound. But it's pretty darn difficult to devise one that will do equally well flying both slower *and* faster than the speed of sound.

Designing organizations for tomorrow is just as difficult. We will have to come to understand that the way we built and grew companies in the "subsonic" age won't work anymore. In effect, we will have to learn the aerodynamics of the era of networks, and understand how to build organizations that can thrive in a world where speeds are reversed.

CHAPTER 3

LINEARITY
IS DEAD

We won't be able to understand and leverage
the age of networks if we don't start
thinking about our world in a new way—
start thinking of our businesses,
our companies and our markets as complex,
internetworked, adaptive systems.

Everyone loves simplicity, but when it comes to understanding your market (or your world), it just doesn't work anymore. Learn why complex, ever-changing networks are the emerging language of today—and how to speak that language's core vocabulary.

This is a vital chapter of the book. What lies ahead may seem daunting at first, because it involves mathematics, some physics, and even a little metaphysics. But I promise it will be easy to read, and you'll get to know some pretty interesting characters along the way. And if you get through this chapter, the rest will be smooth sailing.

Readers of my last book said that they really liked the chapter about the *limits* of the digital era. I was told that the mathematical equations made readers look really smart to their friends. Well, wait until you get a load of this chapter. You can look forward to impressing your friends at cocktail parties with mentions of Ilya Prigogine, complex adaptive systems, and Schrödinger's wave function. You'll be the life of the party!

But let me cut to the chase: simple doesn't cut it in today's society. Although we would like things to be simple and linear, the world is different now. In this chapter, we'll explore the differences among linear systems, dynamic systems, and complex systems.

The dynamics in the era of networks seem to be very close to the world of complex adaptive systems, which resemble many of the structures and organisms found in nature. The key to understanding complex adaptive networks is to focus on understanding the nature of the connections, the strength of the relationships, and the intensity of the information flow in a network.

Only then can you innovate at the edge of chaos.

CHASING SIMPLICITY

As humans, we strive for simplicity. We want our lives to be simple and clean. We want our relationships with our loved ones to be easy and straightforward. We strive to make our organizations, rules, laws, structures, economies, and governments simple to understand and simple to deal with.

But they never are.

Most of us chase simplicity for our entire lives. We think our lives would be so much better if we just had empty e-mail inboxes. We think we'd be happier if we didn't have cluttered garages filled with boxes of stuff that we'll never use again. And we think we could achieve total inner peace and tranquility if we could just, "empty our minds."

But we can't.

Life is not simple. Life is not clean. Life is entirely messy, utterly complex, and above all nonlinear. And that is exactly what I want to explore in this chapter: how we must abandon the concept of linearity when we are thinking about structures and organizations, about markets and customers.

Linearity is dead.

LINEAR SYSTEMS

The reason we crave linearity is that it is simple. Linear systems are easy to understand and extremely easy to control. We understand linear systems intuitively.

We can switch a lightbulb on or off. Simple. We can turn the water in our showers on, and we can get more water if we turn the knob farther to the right. Simple. We can change the temperature by turning the dial, and get more or less heat based on our simple linear control.

But unfortunately, the majority of the systems that surround us are not that simple. Most of them are essentially nonlinear systems. And although we can understand the rules that govern them, we can't always pinpoint cause and effect in nonlinear systems.

One of best illustrations of this is the dual pendulum. If we take a simple pendulum and let it swing by the simple force of gravity, we understand the mechanics of this movement intuitively. The pendulum will swing from left to right, with maximum velocity at its lowest point. It will gradually climb until it halts, then swing the other way. It's simple to understand, and it's very predictable.

Now let's change the pendulum from a one-legged to a two-legged pendulum, connected by a hinge. When we let it swing now, we get a pattern that seems completely unpredictable and totally erratic.

The one-legged pendulum is a simple linear system, governed by the laws of gravity. The two-legged pendulum is a dynamic system, also governed by the exact same laws of gravity. But the fact that the two legs of the pendulum are connected makes this a much more complex system to describe.

As a matter of fact, describing the movement of a two-legged pendulum properly requires advanced mathematics. Throughout history, this is what has happened time after time. Whenever we dig a little deeper into the way our world works, we seem to need a new language to describe it.

DYNAMIC SYSTEMS

When we have systems that act on each other because they are connected or because they influence each other, we have to introduce the concept of feedback. Feedback is the fundamental characteristic of dynamic systems, and it often introduces a complexity that requires us to use a different mathematical language to describe how it works: calculus.

Now bear with me. I know some of you are scared of the "c word," but it's not new. We've known about equations and calculus for centuries. The problem is that most people don't really like mathematics, and that most corporate executives faint when they hear the phrase *differential equation*.

Let me offer a little historical context. At the end of the seventeenth century, the world was evolving at a breakneck pace. Armies were using new technology to fire bigger and bigger cannonballs, and they needed to have precise mechanisms to calculate the trajectories of these cannonballs. Fleets of ships were navigating the globe, and in order to have better navigational insights, they needed more accurate timekeeping devices and better methods of positioning. All this required mathematics that went beyond the simple rules of algebra, paving the way for calculus.

Two men laid claim to inventing modern calculus at basically the same time. The German mathematician Gottfried Leibniz and the English physicist Isaac Newton fought a bitter duel over who invented modern calculus. Leibniz started working on the development of calculus in 1674, and was the first to publish a paper using it in 1684. Newton claimed that he had already invented it in 1666, when he was 23 years old, but that he had not published his works.

Today, we're almost certain that Newton was in fact the true inventor. But the most interesting thing was the reason why he invented it: he wanted to understand the movements of the planets. He invented a whole new mathematical language to explain exactly why the planets moved the way they did.

Newton developed three laws of motion and a law of universal gravitation. These laws changed astronomy forever. But maybe even more important, he invented a new language that gave us a hold on dynamic systems.[1]

The beauty of calculus is that it gives us a mathematical way to describe how nonlinear systems work. We can describe how the two-legged pendulum operates, how weather systems work, and how electromagnetism and electricity function. But that doesn't make these things simple. And although calculus gave us an elegant mathematical formulation of nonlinear systems, we're still lost on the simple cause-and-effect control mechanisms.

Newton's breakthrough in mathematics unleashed a way to discover the world as never before. After the development of calculus, science and technology grew by leaps and bounds. The Industrial Revolution came about because we had mastered the language of these dynamic systems.

The worlds of chemistry, materials, and electricity opened up. We could now build canals, bridges, machines, and eventually the Eiffel Tower.

The progress of science that fueled the Industrial Revolution also triggered the need for a place to showcase the new wonders of progress, science, and technology. Today we'd build a website, grow a social network, or develop a YouTube channel. But since those didn't exist during the Industrial Revolution, we settled for world's fairs.

Two exhibitions were absolutely crucial during this period: the Paris Exhibition of 1889, with the Eiffel Tower as its centerpiece, and the Chicago Columbian Exposition of 1893.

The Paris Exhibition was held during the year of the 100th anniversary of the storming of the Bastille, the symbol of the beginning of the French Revolution. The Expo was a huge success, attracting more than 32 million visitors. And to show off France's great engineering expertise, the Eiffel Tower was constructed of wrought iron and served as the entrance arch to the fair.

The tower was initially designed by two French engineers, Maurice Koechlin and Émile Nouguier, who worked for the construction company of Gustaf Eiffel. In the beginning, Eiffel himself showed little enthusiasm for the ideas of his two engineers. But when the plans for the World Exhibition started taking place, Eiffel realized the potential of the monument and took charge of the project. He bought the rights to the patents on the tower's design and established a separate company to build and operate it. It would become a monument to the progress of mankind in the industrial age.

But the tower was not just a symbol of a glorious exhibition and the power of the French Republic. It was also a symbol of the advancement of science. Gustaf Eiffel had the names of 72 French scientists, chemists, astronomers, mathematicians, and engineers engraved on the sides of the tower just below the first balcony. The statement was clear: this is the age of technology.[2]

After the Paris Expo, it was America's turn to create a world's fair, one that would be the "fair to end all fairs." It was to be called the Chicago Columbian Exposition, to honor the 400th anniversary of Columbus's voyage.

NEW LANGUAGE OF SYSTEMS

Just like the exposition in Paris, the Columbian Exposition was also a celebration of technology. "The electrical hall was filled with telegraphs and telephones, electric railways, elevators and lighting." As a matter of fact, the exposition became the battleground for one of the fiercest fights in the history of technology. Two giants, Thomas Edison and Nikola Tesla, went head to head. At stake was what the electric technology of the future should be: AC or DC?[3]

Today we find it completely normal to plug our vacuum cleaner into a wall socket powered with 220 or 110 volts of alternating-current electricity. But in 1893, the battle between the two competing technologies was on. For the exposition, Thomas Edison's General Electric Company proposed to power the electrical exhibits with direct current at a cost of almost $2 million. Westinghouse—with the help of Nikola Tesla—proposed using its alternating-current system to illuminate the Columbian Exposition for $399,000. Westinghouse won the bid.

Tesla was the star of the show. He wowed the crowds with a series of electrical effects, using high-voltage, high-frequency alternating current to light a wireless lamp and shoot lightning from his fingertips. A Serbian-born American inventor, he pioneered the idea of alternating current. He was convinced that AC was a much more efficient way to transport electric energy over long distances. Edison didn't believe it, but he didn't understand the mathematics that Tesla had mastered. He didn't understand the new mathematical language.

Edison turned out to be a sore loser. He portrayed AC as dangerous, and he engaged in an elaborate public relations campaign to debunk it, discredit Nikola Tesla, and ruin George Westinghouse. He even had a professor named Harold Brown go around talking to audiences about the dangers of AC, electrocuting dogs and old horses on stage to show how dangerous it was.[4]

Here's the point: we need to understand the new language of the systems we're dealing with. If we don't, we won't understand the potential of a new technology. We won't be able to observe the world in new ways. Thomas Edison, the "Wizard of Menlo Park," was a true pioneer, entrepreneur, and genius. But when it came to the future of AC, he was on the wrong side of history. Without the right understanding of the underlying rules of technology, his gut feeling had him completely fooled.

COMPLEX ADAPTIVE SYSTEMS

Remember how erratic the movement of the dual pendulum became when the system had just two elements and a hinge? Imagine how much more erratic the movement would be if you had hundreds, thousands, or millions of agents, all acting and reacting upon one another, and all providing feedback. Welcome to the world of complex systems.

Complex systems are all around us. Pretty much everything we observe in nature consists of a complex system with a very large number of agents. Take, for example, the air and water molecules in a weather system, or the flora and fauna in an ecosystem. Look at the economy, the human brain, developing embryos, or ant colonies. All of these are examples of complex systems. Each of them is based on a network of many agents acting in parallel. In a brain, the agents would be nerve cells. In an ecosystem, the agents would be the different species. In an embryo, the agents would be cells, and in an economy, the agents would be individuals, corporations, or households.

Weather is another classic example. In a weather system, the number of agents—air and water molecules—is so vast that we can't even begin to describe the individual relationships in mathematical language anymore. We understand the mathematics that governs the forces on the molecules, but the problem is that there are so many agents. Thus, we have to observe the system in a completely different way, and again in a completely new language.

The key is that instead of focusing on the individual agents, we have to observe them as a whole. In a complex system, all these agents will interact and connect with one another in unpredictable and unplanned ways. But from this mass of interactions, regularities will emerge and start to form a pattern that feeds back on the system and informs the interactions of the agents. The key word here is *pattern*. We can't describe the exact state of one particular agent, such as one particular water molecule in a tornado. Instead, we focus on the emergence of patterns, like the different cloud formations. As writer Peter Fryer puts it, "For many years, scientists saw the universe as a linear place. One where simple rules of cause and effect apply. They viewed the universe as a big machine and thought that if they took the machine apart and understood the parts, then they would understand the whole." But despite using the most powerful computers in the world, the weather remained unpredictable; despite intensive study and analysis, ecosystems and immune systems did not behave as expected.

Gradually a new theory emerged—*complexity theory*—based on relationships and patterns. Complexity theory maintains that the universe is full of systems that are both *complex* and constantly *adapting* to their environment. Hence, they are complex adaptive systems.

ENTROPY

The first time we started to think about these systems, composed of an immense number of smaller agents, it actually gave rise to a completely new language in the world of physics. The result was the branch of science we call *thermodynamics*. The word *thermo* is derived from the Greek word *therme*, or heat. That's exactly where the field originated: trying to study the dynamic effects of the heat used in steam engines, the central technology of the Industrial Revolution.

One of the pioneers of the field was the French physicist Nicolas Léonard Sadi Carnot, who believed that the efficiency of heat engines could help France win the Napoleonic Wars. Sadi was very much interested in trying to improve the performance of steam engines, and he wrote his magnum opus, *Reflections on the Motive Power of Fire*, in 1824.[5] It became the founding scientific basis for the discipline of thermodynamics. In it, he described what would later be known as the second law of thermodynamics: heat cannot spontaneously flow from a colder location to a hotter one.

But there are two fundamental issues here. The first is the concept of entropy. Entropy is a measure of the disorder in a system containing energy or information. The less ordered a system is, the greater its entropy. A general characteristic of natural processes is that they lead to an increase in entropy—that is, to greater disorder.

The second fundamental issue is that this process is irreversible. Most laws in physics don't have a direction. They're usually very symmetrical, but this one is not. Entropy increases over time, and you can't reverse this without outside intervention. All complex natural processes are irreversible.

What the field of thermodynamics opened up was the fact that in order to describe the world as we know it, and in order to describe the reality of natural phenomena and processes, we have to abandon the concept of understanding exactly where every individual element is. Instead, we have to look at the bigger picture, understand the patterns, and use the language of statistics and probability to describe the behavior of natural processes. We have to abandon the absolute and embrace the realm of probabilities.

SCHRÖDINGER'S CAT

This new way of looking at the world came to full fruition with the advent of quantum mechanics in the early twentieth century. For many years, scientists had found that when you applied the laws of physics to the extremely miniature world of atoms and subatomic particles, the laws just didn't seem to make sense.

The real genius behind the completely new language and field of quantum mechanics was the Austrian physicist Erwin Schrödinger. He won the Nobel Prize in Physics in 1933 for the introduction of his wave equation, which forms the heart of quantum mechanics. He will be most remembered, however, for his cat, Milton.

He introduced his famous "Schrödinger's cat" paradox in a 1935 paper. In it, he tells the story of a cat sharing a closed box with an elaborate booby trap mechanism consisting of a vial of cyanide gas, a small quantity of radioactive material, and a radiation detector. If the radiation detector senses decay in the radioactive material (a random, 50/50 proposition), it triggers the release of the poison gas, and the cat is killed. But if radioactive decay is not detected, then the cat enjoys a quiet nap and no harm is done.

So long as the box remains closed, scientists cannot observe whether or not the cat is dead. So until the box is opened and the cat is observed, the cat exists in an indeterminate state and must be assumed to be both dead and alive. This paradox highlights some of the absurdities that arise when we abandon the concept of the absolute and have to reason in the realm of probabilities.

DEPRESSED YET?

So far in this chapter, we've abandoned the concept of linear thinking and given up on the concept of absolute truth. What else? Well, we also now know from the second law of thermodynamics that the sad fate of the universe will be that entropy will keep increasing until all order has disintegrated into chaos. Great.

This is the point where I have to start figuring out a way to nurse you back to mental health by the end of this chapter. Enter one of the most intriguing of all characters: Ilya Prigogine.

Ilya Romanovich Prigogine was born in Moscow just months before the Russian Revolution of 1917. His father was a prominent chemical engineer at the Imperial Moscow Technical School, and his mother was a pianist. The family fled Russia in 1921, first to Germany and then to Belgium. He became a professor at the Free University of Brussels, and in 1977 he was awarded the Nobel Prize in Chemistry for his work on *dissipative structures*.

Prigogine discovered that when he was studying these structures, some of them did not follow the standard route to entropic death and fall into a state of disorder. He showed that when a physical or chemical system is pushed away from equilibrium, it survives and thrives, whereas if it remains at equilibrium, it dies. The reason is that when they are far from equilibrium, systems are forced to explore. This exploration helps them create new patterns of relationships and different structures. The second law of thermodynamics states that in any isolated physical system, order inevitably dissolves into decay. Prigogine showed that in a system that is powered by an energy source—the Earth bathed in light and heat from the Sun, for example—structures could evolve, become more complex, and thrive.

Computer simulations of complex, evolving systems also demonstrate that it is possible for the order of new survival strategies to emerge from disorder through a process of spontaneous self-organization. It now seems that self-organization is an inherent property of a complex adaptive system.[6]

THE AGE OF NETWORKS

By this time, you're probably beginning to wonder what this book is all about. It is my fundamental belief that we are entering an era that will be known as the age of networks. All around us, thanks to the New Normal effects of digitization, we've started to build a society that is entirely based on the concept of networks. We have networks of information, networks of knowledge, networks of entertainment, networks of friends, and networks of enterprises. Everything we see around us is based on the concept of networks.

But we still treat our environment, our economies, our markets, and our organizations as simple linear systems. We still hope for simple cause-and-effect relationships when we launch a marketing campaign to convince customers of the merits of a product or service. We still hope for predictable outcomes when we launch a reorganization of a corporate enterprise, hoping to make it more effective in the marketplace.

It is my deep conviction that we won't be able to understand and leverage the age of networks if we don't start to think about our world in a new way—start thinking about our businesses, our companies, and our markets as complex, internetworked, adaptive systems. Just as with every new paradigm of insight in the world of physics, every new understanding of how the world actually works, we have to understand the new language of the age of networks. That language is based on the behavior of complex adaptive systems.

THE CORE VOCABULARY OF COMPLEX SYSTEMS

All right, there is a new language to be learned, one that will help us make sense of our new era of networks. So, what's the core vocabulary that we need to master in order to speak the new language?

In the *Hitchhiker's Guide to the Era of Networks*,[7] this would be the core vocabulary list that would allow you to get by and survive—the absolute minimum you have to know if you are to understand the world of complex systems:

1. Complex systems exhibit *emergence*. Complex systems don't have a master plan—they evolve, they emerge. From the interaction of the individual components of a system comes some kind of global property, or pattern—something that you could not have predicted from what you know about the component parts.

    For example, an anthill is a wondrous piece of architecture, with a maze of interconnecting passages, large caverns, ventilation tunnels, and much more. Yet there is no grand plan, no grand design. The hill just emerges, the result of the ants following a few simple local rules. This emergence is a key phenomenon of complex systems.

2. Complex systems are built on *connectivity*. Complexity arises from the interrelationship, interaction, and interconnectivity of all the elements within a system and between a system and its environment. This means

that a decision or action by one element within a system will affect all other related elements, but not in any uniform way. It's the old dual pendulum on a massive scale.

The ways in which the agents in a system connect and relate to one another is critical to the survival of the system, because it is from these connections that the patterns are formed and the feedback is disseminated. The relationships among the agents are generally more important than the agents themselves.

When we look at networks today—the degree of interconnectedness, the way a network is formed, the way the nodes and connections within a network grow—it is vital that we understand the patterns that can emerge.

3. Complex systems evolve in *coevolution*. A complex system exists within an environment, but it is part of that environment. Changes in a complex system have to be seen as coevolution, instead of mere adaptation to the environment. As its environment changes, the system needs to change to ensure the best fit. But because the system is part of its environment, when it changes, it changes its environment, thereby requiring the system to change again. It's a constant process.

4. Complex systems are *not perfect*. They are suboptimal. A complex adaptive system does not have to be perfect in order to thrive within its environment. It only has to be slightly better than its competitors. Therefore, complex systems are not designed for optimal efficiency, but are based on the concept of *perfect enough*.

5. Complex systems favor *variety and diversity*. The greater the amount of variety and diversity within the system, the stronger that system can become. The key to a complex system's evolution is its ability to explore new possibilities, new ways of coevolving with its environment. This means that the strength of a complex system is based on the variety and diversity of combinations, perspectives, or compositions that it contains.

6. Complex systems are *self-organizing*. There is no hierarchy of command and control in a complex adaptive system. There is no planning or managing, but there is a constant reorganizing to find the best fit with the environment.

Complex systems evolve on the edge of chaos. That is the true legacy of Ilya Prigogine.

A system that is in equilibrium lacks the internal dynamics to respond to its environment and will slowly (or quickly) die. A system that is in chaos ceases to function as a system. The most productive state seems to be at the edge of chaos. That's where there is maximum variety and creativity, leading to new possibilities and the best chance for survival.

SCALE-FREE NETWORKS

The final part of this chapter is dedicated to the latest thinking in this field—the introduction of scale-free networks. We're almost there, but it will involve a trip to Transylvania as well as a quest for Kevin Bacon.

When we talked about complex systems, we used the word *agent*, and we have mostly assumed that all agents are created equal. But in the real world, that might not be the case.

In many of the networks that rule the age of networks, that is not exactly true. In these cases, we see that some agents are more special, more influential, and more important than others.

The chief scientist who discovered the properties of these types of networks is Albert-László Barabási, who was born in 1967 in Transylvania. He studied physics at the University of Bucharest. Later he worked briefly at IBM, and then in 2004 he founded the Center for Complex Network Research.

It was Barabási who applied the concept of complex adaptive systems to the world of networks. He found that in the age of networks, the most interesting networks were the ones that he termed *scale-free networks*. These networks don't have purely random agents that are randomly connected. Instead, they have hubs that are crucial to how these networks grow and evolve.

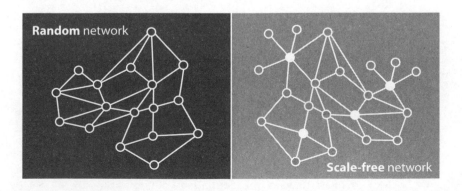

So what about Kevin Bacon?

Actually, Barabási has a Kevin Bacon number of 1—both appeared in *Connected*, a documentary about complex networks. The notion, of course, comes from the "six degrees of separation" concept, which states that any two people on Earth are six or fewer acquaintance links apart. That idea eventually morphed into "Six Degrees of Kevin Bacon," the game in which almost any Hollywood star can be linked to Kevin Bacon in six movies or less. So how does this apply to what you're reading about? Well, some of the nodes on the Internet are more active, more influential, and more important than others. That's why the most highly connected nodes have been referred to as the "Kevin Bacons" of the Web.

In his study of scale-free networks, Barabási showed that we can apply the notion of complex adaptive systems to the age of our networked society. As he remarked, "We caught a glimpse of a new and unsuspected order within networks, one that displayed an uncommon beauty and coherence."

CONCLUSIONS

After reading this chapter, I sincerely hope that you aren't suffering from Hellenologophobia, or the irrational fear of Greek terms and complex scientific terminology. If you are, I apologize for introducing you to the world of differential calculus, entropy, thermodynamic cycles, quantum wave functions, and scale-free Barabási networks.

But seriously, we're entering an age of networks right now, where everything we do, everything we know, and everything we aspire to is based on a new concept. It's an infrastructure that's like nothing our society has ever experienced before.

Newton said, "If I have seen further, it is by standing on the shoulders of giants." By this, he meant that his achievements and insights were made possible by the myriad of scientists who had come before him, puzzling their way through the mysteries of the universe.

But today's world is different.

In today's networked society, the concept of standing on the shoulders of giants has been replaced by that of connecting through the nodes of the network. Networks are our future. Networks are all around us. But remember, networks are complex systems. They are evolving, complex, adaptive systems.

And therefore, if we want to understand the future, we have to understand a new language: the language of networks.

May you learn to speak the language of networks fluently.

HOW THE MEDIA
DISCOVERED NETWORKS

THE GOLDEN AGE OF WIRELESS

The media industry has understood the power of networks for a very long time. But today, the full power of networks is beginning to shift the power base of the media industry more forcefully than ever before.

When we think of networks in the field of media, many of us think of the big, powerful television networks—familiar names like ABC and NBC in the United States or the BBC in England. But truth be told, those organizations are probably the old-school networks. They're more accurately described as broadcasters, because that's exactly what they do: they broadcast content to a huge audience, whose members all enjoy the same content at the same time.

In the age of the New Normal, that sounds pretty old-fashioned. But 50 years ago, this was novel and groundbreaking stuff. The grandfather of these television and radio networks was a man by the name of David Sarnoff. And his story is truly remarkable.

Sarnoff was the founder of RCA (the Radio Corporation of America) and the leading man behind the introduction of radio and television to the world. No other man has singlehandedly transformed media as much as this emigrant from the tiny village of Uzlian, in the province of Minsk in Belarus.

David Sarnoff immigrated to the United States at the age of nine. His father, Abraham, had already been in the United States for a few years, and had saved enough to bring his entire family to New York. The poor young Russian family faced hard times when they arrived. David got a job as a newspaper courier, pocketing a quarter for every 50 newspapers that he sold. He soon realized that he needed to move on to bigger things and set up his own newsstand in Hell's Kitchen, on the lower west side of New York City.

Sarnoff became obsessed with the idea of communications and came to acutely understand the power of information. He saw firsthand the hunger for content by observing the consumers of his newspapers. After his father became ill, he needed to provide more money for the family, so he decided to take an extra job as a messenger for the Commercial Cable Company, a company that handled cable traffic across the Atlantic. He watched the telegraph operators handling messages in Morse code, and he taught himself the art of Morse telegraphy. David Sarnoff eventually left that job and landed at the Marconi Wireless Telegraph Company. And then the magical moment of his life happened.

In December of 1906, company founder Guglielmo Marconi walked through the door into his New York office. An eager David Sarnoff was waiting for him, and he managed to impress the Italian CEO so much that he became Marconi's personal assistant. With this newfound partnership, everything changed.

A seminal moment came on April 14, 1912. Sarnoff was on duty when the RMS *Titanic* sent out a dreadful message that has been in the history books ever since: "S.S. *Titanic* ran into iceberg. Sinking fast." The unsinkable *Titanic* had run into an iceberg in the waters of the North Atlantic Ocean. Sarnoff heard the message from the doomed ship.

Equipped with the Marconi Wireless System, Sarnoff worked around the clock for days. He received the lists of people who had perished and those who had survived. More than 1,500 people lost their lives in the disaster, many of them from prominent and wealthy families, and their loved ones stood outside the Marconi station waiting to hear news from Sarnoff. His diligent work in the face of tragedy made him a hero.

But he also came to realize something profound that day. As he was shouting out the names of survivors, he discovered the true power of broadcasting information. According to Todd Buchholz, a former White House director of economic policy, "The *Titanic* 'made' wireless, and, incidentally, helped make Sarnoff too."

After the incident, Sarnoff became obsessed with delivering sound to as many people as possible. He wanted to build a system that wouldn't provide mere one-to-one messaging, as in one telephone operator to another, but instead would send messages to a network of people. He wrote a memorandum in 1916 outlining his ideas for making radio a household utility, bringing music into houses by means of wireless, and the conception of a "radio music box."

The Golden Age of Wireless was born.

As the Roaring Twenties rolled along, radio became immensely popular, and sales of radios boomed. RCA gave birth to two new networks: ABC and NBC. In 1929, engineers figured out how to put radios in automobiles, and a whole generation could listen to music and news while driving.

In 1929, Sarnoff became the president of RCA, just before the stock market collapsed and the Great Depression began to set in. But he did not slow down. On the contrary: envisaging a world beyond radio, he wanted to broadcast pictures to the world.

Sarnoff was no scientist; he was a visionary and a businessman. He gave the technical challenge to his chief scientist, Vladimir Zworykin. At the 1939 New York World's Fair, they were ready. They broadcast the address of President Franklin Roosevelt to approximately one hundred television sets, which probably gave it an audience of 3,000 viewers. It was a pretty meager start. But Sarnoff wanted to push even further. He envisioned color TV.

I'm old enough to remember when my family had a black-and-white television set. And I remember the thrill it was when color TV entered our household. David Sarnoff brilliantly created a demand for his new product by commissioning Walt Disney to develop a show called *Walt Disney's Wonderful World of Color*. I mean, who would want to watch that show on a black-and-white TV?

David Sarnoff was a true pioneer who understood the power of mass communications—the power of broadcasting information to a vast audience. But his notion of networks is old-school now.[1]

YOUTUBE AND THE NEW NETWORK

Today, we see the media industry being transformed by the power of networks, real networks this time. This is no longer the broadcasting era, where content dictated by a few was distributed to many. This is the era where the power of information flows directly—where content, audio, and video are distributed, rated, and recommended by the nodes of the network.

Take YouTube, for example. YouTube is impressive by any standard. Hundreds of millions of viewers watch content on YouTube every day, and millions more comment and share the videos they like. YouTube is now the second most popular search engine in the world. And if you still think YouTube is primarily videos of cats on skateboards, you're dead wrong. (But yes, there are a lot of videos of cats on skateboards.)

YouTube has opened a brand-new, state-of-the-art studio in Los Angeles, where it invites people to come and tell their stories on—where else?— YouTube. Anyone—well, almost anyone—can come in and use the top-notch facilities for free to produce content on YouTube. I had the chance to visit the place, and it is awesome.

In the hallway of the YouTube Space studios stand two vintage arcade games, where visitors and employees can play Pac-Man or Asteroids. They're wonderful to play, but vintage they are. They're relics of a past in which video games were custom-built, expensive machines that were found in arcades, only to one day be replaced by Xbox and PlayStation. Eventually these gave rise to a whole generation playing Angry Birds on their smartphones.

YouTube couldn't have picked a better attribute to show in its new studios. It couldn't have picked a better location, either: the new studios are housed in the old Hughes Helicopters factory. Actually, one of Howard Hughes's helicopter models is proudly parked right out in front of the building.

I have two kids, aged 11 and 15. It's safe to say that in the New Normal, where digital has become a normality, I learn more from them about all things digital than they learn from me. I notice that my teenage daughter is now spending less and less time watching daytime TV. There are still the occasional moments where she slouches in front of the television set and takes in *iCarly* episodes or *SpongeBob SquarePants*. But those moments are fewer and farther between.

Instead, she's spending more and more time watching content on YouTube. At first, she searched YouTube for the stuff she knew from television. But then something flipped, and she quickly latched onto channels and series that exist only on YouTube.

Today, my daughter is hooked on watching Smosh. I can't stand it. I probably have the same look on my face when I watch Smosh as my folks had when I was watching episodes of *Dr. Who* at her age. My father never got the Daleks. And I don't get Smosh.

Today, Smosh has more than 12 million subscribers, and more than 2.6 billion video views. That's right, *billion*.

Smosh is on YouTube only. Its creators don't want to be on TV, because, "We are where our audience is. And our audience is on YouTube, not on plain old TV anymore." In other words, they think regular TV is dead as a doornail. And they may be right.

Smosh is not a unique phenomenon. There are thousands of successful channels on YouTube. Some of them are incredibly successful. YouTube is producing stars by the dozen. Take Michelle Phan, a makeup instructor who has a channel where she shows makeup tips, talks about beauty products, and has millions of subscribers. These new YouTube stars all have one thing in common: they're twentysomething digital natives who have always known things to be digital. They don't want to be on TV.

It's pretty clear that the Oprah and Jay Leno of the next generation won't be coming from TV. They'll come straight from YouTube. That's why YouTube built a studio in Los Angeles. And one in London, and another in Tokyo. But the Los Angeles facility is its flagship studio. Where Howard Hughes once built helicopters, the next generation of talent is producing the next YouTube hit, the next monster YouTube channel, the next generation of content.

We're witnessing the end of an era. Television won't go black soon. But watching linear television as we know it could become the pastime of those who will sit in front of their television after they've read their old analog paper newspapers. Let's face it: why would we watch old-school television when we have the forever streaming libraries of Hulu, Netflix, Amazon Prime, and Aereo where we can see what we want to at the moment we want to? Or when we can purchase content from iTunes, Amazon Instant, and Google Play? Why indeed?

David Sarnoff saw the power of mass communications. Today we're seeing the power of networks become the dominant new factor in media. It's reshaping media right in front of our very eyes. In media, it's pretty clear: networks will win.

CHAPTER 4

INFORMATION BECOMES A FLOW

In the past we viewed information
as static, like water in a reservoir or a pond.
All of a sudden, we have entered the era of networks,
and information has started to flow.

Stagnant pools of information are a thing of the past. Today's information flows like a river. To ride those rapids and thrive in the age of networks, you'll need to understand how and why information flows, both within and outside your organization.

IT'S A SMALL, SMALL WORLD

During the 1960s, the United States was coming into its own. A global superpower that was at its peak, the United States became the driving engine of the Golden Age of Consumerism. Big corporations were becoming the new leading frameworks of power. It was the dawn of the Age of Technology, the Space Age, the Atomic Age, and the advent of the Age of Computers. And New York was at the center of it all.

In 1964, New York played host to the World Exposition. The event was both a rousing success and a colossal financial failure. The problem with the 1964 edition was that it wasn't recognized as a "real" world's fair. A group of New York businessmen had conceived the idea of having a world's fair in their city and had hired Robert Moses to run the corporation established to govern it. Moses selected Flushing Meadows Park, a converted Queens garbage dump, as the project's location. It was to be a grand jamboree, a celebration of the greatness of the United States and the city of New York.

There is an international body in Paris called the Bureau of International Expositions (BIE). Its main purpose in life is to sanction world's fairs. When the bureau heard of the preposterously commercial plans for the New York fair, it summoned Robert Moses to Paris. Moses was a powerful figure in New York, but the people in Paris were not impressed by his bravado. When he used the press instead of the customary quiet diplomacy to argue his case, the BIE responded viciously by formally requesting that all of its member nations not participate in the New York fair.

This nearly destroyed the project. Great nations such as Canada, Australia, and the Soviet Union would not be present and, more important, would not send their masses to the fair. This left the fair to smaller countries, which came in droves (the Vatican had an extremely popular pavilion). But more important, it gave unprecedented exposure to big corporations.

Ford Motors introduced the iconic Mustang to the world. The Westinghouse Corporation planted a time capsule, to be opened in the year 6939. But the biggest winner of the 1964 World's Fair was none other than Walt Disney.

Disney was the undisputed grand master, with Walt Disney Productions designing and creating no less than four shows at the fair. Anyone who's ever been to a Disney theme park has had "It's a Small World" stuck in their head; the song was first exposed to the public at the World's Fair. As well as being a catchy tune, it gave Disney an opportunity to introduce his Audio-Animatronics to the world, in which a combination of electromechanics and computers controls the movements of lifelike robots to act out scenes.[1]

The success of the event galvanized Walt Disney to build something more lasting than the World's Fair. He decided to make plans for a permanent community designed to look at the future, a perpetual showcase of tomorrow. It would be called Epcot, which stands for Experimental Prototype Community of Tomorrow. His vision was of a community of tomorrow that would never be completed, but would always be introducing, testing, and demonstrating new materials and new systems. Sadly, Disney died in 1966, before his vision became reality.

FLOW OF MAGIC

I visited Epcot as a child, and I thought it was truly amazing. The combination of a technology showcase with the atmosphere of a permanent World Exposition was thrilling to me as a youngster.

I recently went back with my children, and they found it to be less than exciting. True, it's not easy to keep "new technology" fresh for 30 years, and many of the exhibits and attractions seem eerily faded and dated. Some of them actually look more retro chic than futuristic.

But the most interesting thing about the experience was the use of the MagicBands. That part, my kids loved. Introduced at Disney near the end of 2013, MagicBands are wristbands containing a computer chip that can be used for all sorts of functions. You can open your Disney hotel door, order a drink at the pool, enter the park, pay for your meal, or buy a souvenir at the ubiquitous gift shops.

MagicBands are extremely easy to use. You just hold your wristband against the Mickey-shaped receptors at the rides, shops, and restaurants. For the user, it's an extremely convenient all-in-one device that lets you travel lighter. No more keys, wallets, credit cards, or cash to carry around. And for Disney, it's an information gold mine.

Every individual who enters Disney with a MagicBand on his wrist can have his behaviors observed. Patterns of consumption and the flow of resort guests provide monumental insight into how visitors to the parks actually experience their vacation. From this vast amount of data, Disney can understand how to optimize park flow, optimize revenue generation in shops and restaurants, and ensure that visitors get more bang for their buck.

This type of information is something that we could not have fathomed only a few years ago. Consider the numbers. All the theme parks of Walt Disney World Resort combined have more than 50 million visitors each year. Disney can now follow the flow of those 50 million visitors, tracking where they go, what they eat, and what attractions they take in. The resulting data flow is vast. It's vaster than vast; it's colossal. It's like the way famous science fiction writer Douglas Adams described space: "Space is big. Really big. You just won't believe how vastly, hugely, mind-bogglingly big it is. I mean, you may think it's a long way down the road to the chemist's, but that's just peanuts to space."

INFORMATION BECOMES A FLOW

In the last couple of years, an avalanche of information has hurtled toward us. The pace of information growth doesn't seem to be slowing down—instead, it's picking up it's pace. But the most important element of this experience is that we are now sharing, relaying, and forwarding information faster than ever before. So the fundamental shift is not that there is so much more information out there (although there is), but that we're pumping the information around faster.

One of my favorite examples of this has been the rise of Foursquare. Founded in New York in 2009, Foursquare allows users to connect with friends and "check in" at venues. Its rise has been meteoric, with more than 2 billion check-ins in its first four years of existence. People check in when they get on the bus, when they arrive at work, when they go for lunch, when they get a coffee at Starbucks, and when they're out clubbing with their friends. In the beginning it seemed silly, but it quickly became a sensible thing to do. People want to share on the network. If you're not generating information flow on the network, you're not visible. If you're not visible, you don't exist.

The resulting flows that appear are mind-boggling. You can now view the living pulse of a city, a town, an area, a country, a store, a mall, or an airport by following the flow of information. The pulsations of information, the cadence and beats of user activity flow, are telling signs of how active, how alive, and how exciting a place can be. It allows us to understand the patterns of society and how to influence them.

In essence, information flow becomes visible on the network.

For years, we tried to capture information in reservoirs. We built databases that could store the data that captured all the information that was relevant to our business and our company. We constructed databases that were essentially information prisons. Today, the situation is completely different. The information flows outside our companies are orders of magnitude larger than the databases and data warehouses we've erected on the inside. And this is just a start.

The shift that is taking place now is from ponds to rivers. In the past, we viewed information as static, like water in a reservoir or pond. All of a sudden, we have entered the era of networks, and information has started to flow. If you want to make sense of the age of networks, you will have to understand how the information flows.

INFORMATION THEORY

The father of the science of information theory was a wiry fellow named Claude Elwood Shannon. He was an extraordinary mathematician and engineer, and probably one of the greatest geniuses of the twentieth century. He was also an incredibly playful chap.

Claude Shannon was born in 1916 and grew up in rural Michigan. A smart kid, he was constantly devising and building electrical contraptions. Among them was a wireless telegraph system to a friend's house more than half a mile away. He attended the University of Michigan and went on to MIT, where he graduated in 1937 with a master's degree thesis titled, "A Symbolic Analysis of Relay and Switching Circuits." In his thesis, he developed the mathematical logic foundation that forms the basic concept that underlies all electronic digital computers.

The *Think* movie that was shown in the IBM pavilion at the 1964 New York World's Fair was basically a visual translation of Shannon's thesis, which was called "possibly the most important, and also the most famous, master's thesis of the century."

But Shannon was just warming up.

In 1940, Shannon became a National Research Fellow at the Institute for Advanced Study in Princeton, New Jersey. There, he had the chance to work with eminent scientists such as John von Neumann, and the opportunity to converse with geniuses such as Albert Einstein and Kurt Gödel. At Princeton, at one point Shannon was giving a lecture to mathematicians when the door in the back of the room opened and in walked Albert Einstein. After a few minutes, Einstein whispered something in the ear of a person in the back of the room and suddenly left. Shannon couldn't wait to end his lecture, and afterward he hurried to the back of the room to find out what the great man had had to say about his lecture. The answer was: Einstein had asked directions to the men's room.

During his Princeton years, Shannon began to combine the input from multiple disciplines and started sculpting the concepts and ideas that would become information theory.

Shannon then decided to leave Princeton and go to Bell Telephone Laboratories, in Murray Hill, New Jersey. It was about as close to scientific heaven as you could find. Seven Nobel Prizes have been awarded for work conducted at Bell Labs. Researchers at Bell Labs have given us the transistor, the laser, radio astronomy, and, indeed, information theory.

Early in 1943, Shannon came into contact with the leading British cryptanalyst and mathematician, Alan Turing. Turing had been instrumental in World War II, when he worked for the Government Code and Cypher School at Bletchley Park, Britain's code-breaking center. Shannon and Turing met at teatime in the cafeteria at Bell Labs. It became clear that the work that Turing had done on cryptography was very closely related to the work that Shannon had done on communication theory, namely, how to extract the right "signal" from the interfering "noise" that surrounds the relevant information signal in communication systems.

THE SIGNAL AND THE NOISE

In 1948, Shannon wrote "A Mathematical Theory of Communication," a paper that is sometimes called the Magna Carta of the digital age. It is the basis of information theory as we know it. What Shannon discovered is that you can send a signal without distortion, even in a noisy conversation. He proposed the idea of converting any kind of data—such as pictures, sounds, or text—to zeroes and ones, which could then be communicated without errors.[2]

This was entirely novel. His vision came in an age when communications were still analog. A phone call relayed the caller's voice as a waveform, roughly analogous to her speech. Different communication lines were needed for voice calls and telegraph messages. Shannon realized that everything, once it was broken into digital bits, could travel over the same pipe. Data would be reduced to bits of information, and information transmission would then be measured in terms of the amount of disorder or randomness the data contains: entropy.

It was actually John von Neumann who gave Shannon the idea of calling the concept on which Shannon had stumbled *entropy*. According to the legend, von Neumann told Shannon, "Nobody really understands what entropy is anyway, so you might as well call what you discovered entropy. Since nobody understands that, they won't challenge you."

In short, Shannon basically invented digital communication, as it is now used by computers, CDs, and cell phones. In addition to communication, fields as diverse as computer science, neurobiology, code breaking, and genetics have all been revolutionized by the application of Shannon's information theory. Without Shannon's work, the Internet as we know It could not have been created.

But despite all that, Shannon remained an extremely humble man. In the 1980s, having been away from the field of information theory for many years, he quietly showed up at an international conference, wanting to listen in on the latest findings in the scientific field that he himself had created. He sat there in the audience with no special accolade until the audience began to buzz that the founder of the field was just sitting there. One attendee said, "It was as if Isaac Newton had showed up at a physics conference." When people realized who he was, they pushed Shannon on stage, where he gave a very short speech, then juggled a few juggling balls. Afterward, attendees lined up to get his autograph.

Juggling was one of his main passions in life, along with riding unicycles. If you had been visiting Bell Labs in the early 1950s, you might have seen Claude Shannon juggling three balls in the air as he rode his unicycle down the lab's corridors. If you think that in our day and age, the campuses of Facebook and Google are unique because the engineers in those places seem to decorate their office spaces with weird artifacts and seem to be occupied in exotic stress-relief activities, their original guru was Claude Shannon.

Shannon designed and built chess-playing, maze-solving, juggling, and mind-reading machines. Juggling was a passion, but also a science for him. He developed a unified theory of juggling that calculates how many balls you can juggle anywhere in the universe, based on the gravity of the planet and the number of hands the alien juggler is using. There was a juggling W. C. Fields mannequin in his office that he had conceived and built himself, and a computer that he nicknamed THROBAC that calculated everything in Roman numerals. There were rocket-powered Frisbees, motorized pogo sticks, a mechanical mouse-in-a-maze, and a flame-throwing trumpet. In a 1990 *Scientific American* interview, Shannon said: "I've always pursued my interests without much regard to financial value or value to the world. I've spent lots of time on totally useless things."

At MIT, where Shannon joined the faculty in 1958, tales were told of uncashed checks languishing in his office. Shannon was quite wealthy, in part because he had been profiting greatly from insightful early stock purchases in local high-tech companies based on his theories. Money supposedly never mattered much to him, although friends say he made a fortune from shrewd investments, and by applying his mathematical theories to the stock market. And to the casinos.

THE GAMBLING CONNECTION

The lure of gambling seems to be a very powerful aphrodisiac for the scientifically minded. Vegas is the place to test the intersection of information theory, game theory, and probability, drawing some of the brightest minds in information thinking.

Shannon was no exception. Besides all his gimmicks, he was also the co-inventor, with the MIT mathematician Edward Thorp, of the first wearable computer. Together they conceived a device that would greatly improve the odds when playing roulette.

Shannon and his wife, Betty, made frequent weekend trips to Las Vegas with the Thorps, using the device, which they had built in Shannon's basement. The men played roulette, and their wives monitored the operation, checking to see whether the casino suspected anything. They actually made a small fortune. Shannon and Thorp also applied the same theory, later known as the Kelly criterion, to the stock market, with even better results.

It was natural that Thorp and Shannon hit it off. They were both brilliant and extremely playful. Thorp wrote the first book to mathematically prove, in 1962, that the house advantage in blackjack could be overcome by card counting.[3]

Thorp had used an IBM 704 at MIT to program the equations needed for his theoretical research on the probabilities of winning at blackjack. He put the computer at MIT through its paces and found a way to improve his odds, especially near the end of a card deck that is not being reshuffled after every deal.

Thorp subsequently decided to test his theory in practice in Reno, Lake Tahoe, and Las Vegas. But he needed starting capital to get going.

Thorp started his applied research with $10,000 in seed funding that he got from Manny Kimmel, a professional gambler and a notable underworld figure with extremely interesting mob connections. Kimmel was the founder of the Kinney Parking Company, a New Jersey parking lot company, which he merged with a funeral home company, Riverside; he later expanded into car rentals, office cleaning, and construction. In 1966, the Kinney Parking Company merged with the National Cleaning Company to form Kinney National, and after an aggressive expansion strategy acquired first the Ashley-Famous talent agency, then Panavision, and then in 1969 Warner Bros. After a financial scandal in the parking division, the nonentertainment assets were spun off again in 1971, and the remaining company was renamed Warner Communications, a precursor to today's Time Warner media empire.

So, the MIT mathematician Ed Thorp, with money from the underworld founder of Time Warner, went to Reno and Lake Tahoe to test the algorithms crunched on the MIT computer at the local blackjack tables. The experimental results proved successful, and his theory was clearly verified, since he won $11,000 in a single weekend. He could have won even more had he not drawn the unwelcome attention of the casino security staff, who threw him out, suspecting foul play. Afterward, when visiting Las Vegas to further test his winning system, Thorp would frequently have to disguise himself with wraparound glasses and false beards. Casinos now shuffle well before the end of the deck as a countermeasure to his methods.

Thorp really considered the whole exercise to be an academic experiment—it was probably the first time in the history of computing that a computer was used as a gambling aid. In addition, Thorp became one of the very few applied mathematicians who has risked physical harm in verifying computer simulations. Thorp later joined the University of California in Irvine to teach mathematics, and became a very successful hedge fund manager.

THE GAMBLER'S FALLACY

There seems to be an interesting connection between mathematics, probability, gaming, information theory, and large financial gains—an explosive and incredibly enticing cocktail.

There is a strange, intoxicating drive that we all feel: to win, to beat the system, to understand the patterns. But most often we fall prey to a false belief, falling into the trap of the *gambler's fallacy*.

The most famous example of the gambler's fallacy occurred in a game of roulette at the Monte Carlo Casino on August 18, 1913. That night, a massive crowd gathered around a roulette table as the wheel stopped at black numbers on 26 consecutive spins. The ball fell in black 26 times in a row. Gamblers that night lost millions of francs betting against black, reasoning incorrectly that the streak was causing an "imbalance" in the randomness of the wheel, and that it had to be followed by a long streak of red. They were wrong, and the casino won—big time.

The casino always wins in the long run, but today we see the advent of modern-day players who use probability and information to allow them to look at insights and patterns in a completely new way. Based on the foundations of information science, this has great potential to help show companies how to make sense of the overload of data and information. But again, it starts with gambling.

THE SIGNAL AND THE NOISE, 2.0

Nate Silver is an unassuming character: quiet, rather shy, and with a nerdy look that is accentuated by his retro glasses. He could easily be mistaken for your typical taciturn IT hacker, or a lowly accountant working in the basement of a finance department. In fact, he was once both, but nowadays he is one of the hottest players in the burgeoning field of data science. He's rumored to be able to see the future, or at least to be able to predict it.

Nathaniel Read "Nate" Silver was born in January of 1978. A math whiz at an early age, he also caught the baseball bug when he was only six. But instead of playing baseball, he wanted to predict the outcome of the baseball games. He did it by using data. The term that is used for the empirical analysis of baseball, especially baseball statistics that measure in-game activity, is *sabermetrics*. This term comes from the acronym SABR, which stands for the Society for American Baseball Research. Nate Silver is a prominent member of the sabermetrics hall of fame.

But his real claim to fame came not in the field of sabermetrics, but in the field of *psephology*. This term refers to the science dealing with the study and scientific analysis of elections. Psephology uses historical voting data, public opinion polls, campaign finance information, and all sorts of statistical data to try to understand and predict the outcome of elections.

In 2007, Silver started becoming obsessed with the psephology of U.S. presidential elections. What he had learned from sabermetrics, he started to apply to the analysis and predictions related to the 2008 U.S. presidential election between Barack Obama and John McCain.

He started publishing his findings under the pseudonym "Poblano," and this later turned into his hugely popular blog *FiveThirtyEight*. In the 2008 election, Silver accurately predicted the winner in 49 out of the 50 states. The only state he missed was Indiana, which went for Barack Obama by one percentage point. Not only did Silver predict the next president, but he also correctly predicted the winner of all 35 U.S. Senate races that year.

In 2010, Silver's ideas were getting mainstream traction. In 2012 and 2013, *FiveThirtyEight* won Webby Awards as the Best Political Blog. His predictive powers were so strong, and the belief in his methods so solid, that at one point the Obama campaign turned to him for guidance.

Silver rejects much of the ideology taught in colleges and universities today regarding statistical methods. He believes in a much broader and more diverse variety of sources than the "clean" approaches to data that are used in classical statistics today. Instead of using "sterile" sets of data, Silver develops his predictions using more "messy" sources. He states that in order to predict an outcome, you first need a sound understanding of how baseball, elections, or other uncertain processes work. You must better understand which measures are reliable and which are not. Then, and only then, can you utilize a statistical tool kit to the best extent.[4]

Silver became the poster boy for big data. His book is the bible for how you can apply the art of mathematical model building, using probability and statistics, to large volumes of information.

THE LITTLE ELEPHANT

The technology that powers the big data movement, enabling it to process large quantities of data, is named after a stuffed toy elephant. The name came from the infant son of the technology's inventor, Doug Cutting: Doug's little boy had a stuffed elephant named Hadoop.

Companies such as eBay, Google, and Yahoo! were beginning to create enormous volumes of data inside their companies from the traces left by their users. The clicks of all the eBay users on the various items for sale created gigantic quantities of records. Similarly, the various user clicks on the search results of Google and Yahoo! created mountains of information. Yet, the state-of-the-art technology was hopeless at making sense of all these data. The technology of databases in the last century had never been conceived for this huge amount of data.

The answer came from the talented engineers at Google. They published an article in 2004 on the topic of MapReduce. MapReduce is a programming model for processing large data sets with a parallel and distributed algorithm. This allowed for the filtering and sorting of large sets of information by essentially distributing the complexity of these tasks to a number of nodes in parallel.[5]

You spread the load of the query over many parallel workers to solve the task. This allows the searches to be quicker and much more scalable. Doug Cutting was a Yahoo! employee when he developed an open-source approach to the technology provided by the Google MapReduce solution. In search of a name, he looked no further than his son's toy elephant.

Since then, the world has gone big data crazy. Companies are realizing that in order to make sense of the avalanche of information that is being created, they need new techniques to gather insights, find patterns, and deduce meaning. A whole new industry has been built on top of Hadoop, along with an army of software companies, solution providers, and consultants ready to introduce big data to the world.

However, big data is not magic. You can't just throw all the data you can find on Obama and Romney into a Hadoop-based solution, sprinkle some pixie dust on it, say the magic words, press a button, and out comes the next president of the United States.

More and more, we are coming to realize that the technology we have at our disposal today is making it possible for us to process more information than ever before, faster than ever before, and in real time. But we also still need a human to make sense of it. It's not all hocus-pocus—we still need to interpret the results correctly.

THE MATHEMATICS OF WAR

Maybe the physical opposite of Nate Silver is Sean Gourley. Sean was originally from New Zealand and doesn't look like a nerd at all. He looks more like a golden surfer boy with the build of a track and field athlete than like a computer scientist. Perhaps that is because he is a golden surfer boy with a track and field background.[6]

Sean grew up in New Zealand and got a Rhodes Scholarship to study physics at Oxford, where he got his PhD. Originally, Sean was doing academic research in nanotechnology, but he soon became fascinated with the world of complex adaptive systems. He began to work on understanding patterns of behavior in large complex systems and operations. His PhD thesis coordinator suggested to Sean that perhaps he could apply his thinking and algorithms to the field of war.

Sean began loading up his laptop with any information on wars that he could find. And luckily for him, wars have been extensively documented. His laptop was beginning to understand patterns in wars from the Hellenistic Wars to the Napoleonic Wars, and was getting very accurate results. But Sean wanted to test his hypotheses. He decided to load up his computer with all the publicly available data he could find on a war that was going on at exactly the time of his research work, the war in Afghanistan.

Soon, Sean not only was able to use his big data algorithm to chart the patterns of the war in Afghanistan, but also started to be able to accurately predict where the next bombing, upheaval, disturbance, or turmoil would be likely to occur. By analyzing raw data on violent incidents in the Middle East, Gourley found a surprisingly strong mathematical relationship linking the fatality and frequency of attacks. He published the results in *Nature* magazine. Two days later, he was awakened by U.S. Secret Service agents. He was questioned for days on his findings, and the agents could not believe the results that Sean had come up with. How was it that more than 700 specialized researchers in Langley, Virginia, had worked on the exact same topics, but had come up with less significant results? It was confounding.[7]

Sean was released, but he realized the power of his technology. He left the world of academia, moved to San Francisco, and raised enough venture capital to start his company, Quid. Quid builds tools for making sense of data, seeking meaning and patterns in information based on the algorithms and mathematics of war that he had developed. Today his customer base is impressive, with the U.S. military and intelligence communities making up a significant part of his customer portfolio.

But the essential philosophy that underpins the Quid's technology, and that is embedded in the thinking that Sean Gourley has developed, is based on one central theme: it is *not* about technology alone. It is the combination of humans and technology that makes magic. Sean Gourley calls this concept *augmented intelligence*. The only way to interpret signals and understand the patterns in large information sources—to truly separate the signal from the noise—is by starting to combine our unique possibilities as human beings with the next generation of technologies that are just beginning to appear.

Big data is not a magical solution that can make patterns in data visible. We still need the expert knowledge of data scientists to make those insights tangible. These scientists will become the rock stars of business in the coming years. They are those who have the elusive combination of data savviness and deep business instinct. There are few of them at the moment, and they are all the more sought-after. That's probably the reason why data science competition platforms like Kaggle are so successful, because organizations lack the talent internally and do not seem to be able to find and hire the right people. If your kid says he wants to be a data scientist, I suggest that you open a bottle of champagne, sit back, and rejoice in his bright future.

GRAPHS RULE THE WORLD

What we're beginning to see appear in society, business, and our social lives is that the world is completely defined by networks. The way we look at the dynamics of things around us is not just by observing a lot of individual, isolated, and discrete items. Instead, it's by looking at the connections among them. Networks rule the world, in every aspect of life and business. But for a long time, our world of technology, our designs of information, and our internal organizations have tried to ignore that, instead favoring a stricter, more structured, and more hierarchical view of life. Once we truly enter into the age of networks, this approach won't work any longer. We'll have to learn to adjust. It will be a challenge, no doubt, but there is good news: we have already figured out the math on that problem. Actually, we did so a long time ago.

In 1735, the Swiss mathematician and physicist Leonhard Euler solved the problem of the Seven Bridges of Königsberg. In Euler's day, the city of Königsberg was an important intellectual and cultural center and the capital of East Prussia. It was the largest city in eastern Germany until it was captured by the Soviet Union near the end of World War II. Today it's part of the Russian Federation and is known as Kaliningrad. But the city may be better known for its historical mathematical riddle. The city of Königsberg was set on both sides of the river Pregel and includes two large islands, which in Euler's day were connected to each other and the mainland by seven bridges.

The question was: Can you walk through the city, crossing each bridge once and only once?

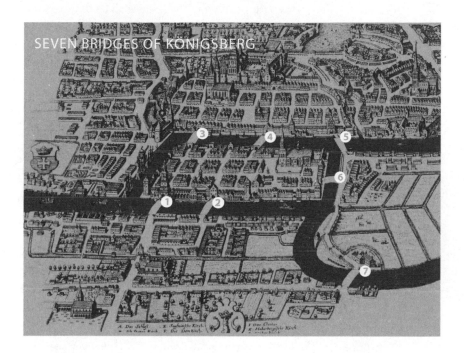

SEVEN BRIDGES OF KÖNIGSBERG

Leonhard Euler proved, unequivocally, that it was impossible. He proved that the problem has no solution, and in his proof, he laid the foundations for the mathematical discipline known as *graph theory*. Graph theory takes into account the topology of a problem: that is, the network structure. In this case, how the river, islands, and bridges were connected was essential to understanding whether the riddle could be solved.[8]

Thanks in part to graph theory, we are now witnessing the early stages of a society that is completely characterized by networks and connections.

Emil Eifrem, the founder of a company called NEO Technology, based in Silicon Valley, built a database technology that was completely based on graph theory. NEO's product is a graph database that allows companies to leverage highly connected data to generate insights and competitive advantages.

Eifrem describes the tremendous impact of graphs as follows: "The first decade of the new millennium has seen several world-changing new businesses spring to life, including Google, Facebook, and Twitter. And there is a common thread among them: they put connected data—graphs—at the center of their business. Facebook, for example, was founded on the idea that while there's value in discrete information about people (their names, what they do, etc.), there's even more value in the relationships between them."

The founder of Facebook, Mark Zuckerberg, indeed built an empire on the insight that he could capture the relationships between the site's members, in what he calls the *social graph*. But businesses in all domains and organizations all around the world are beginning to understand that they are operating in markets that are turning into networks. Companies are waking up to the fact that networks are the dominant structure in business, and that graphs rule the world.

I have not yet encountered many organizations that have adopted the full possibilities of the network. If I had, I wouldn't have had to write this book, of course. But some companies have already evolved rather impressively when it comes to leveraging some of its main characteristics.

Zappos's CEO Tony Hsieh, for instance, announced at the beginning of 2014 that he would transform his entire company—which has a billion dollars of revenue and 1,500 workers—to a much flatter and more collaborative environment, called holacracy.[9] As Hsieh puts it, they are "trying to figure out how to structure Zappos more like a city, and less like a bureaucratic corporation."[10] Cisco, in turn, is investing rather boldly in open innovation with its Entrepreneurs in Residence program. It works with early-stage entrepreneurs with big ideas for enterprise solutions who join its start-up incubation program. Samsung Electronics has turned to crowdsourcing: it announced a worldwide challenge for Gear 2 app developers—the Samsung Gear App Challenge—in order to support the wearable app market.[11] But I'm running ahead of myself here.

THE RICH CLUB

Back in 1897, the Italian economist, sociologist, and philosopher Vilfredo Pareto observed that there was great social and economic disparity among people in different societies and countries. He will remain world-famous for his legendary 80/20 rule, after he realized that 80 percent of the land in Italy was owned by only 20 percent of the population. He was also responsible for popularizing the use of the term *elite* in social analysis.

When we talk about networks in which some nodes have more influence than others, we often use the term *weighted networks*. Within such a network, each node's weight represents its influence, importance, or strength. These nodes might be neurons, individuals, companies, groups, organizations, airports, or even countries, whereas the connections and links can take the form of friendships, communication, collaboration, partnerships, alliances, or information flow.

Pareto showed us that in real life, some elements have more influence than others. The same can be said of the world of graphs and networks. A minority of the nodes in a network can be responsible for a vast majority of the dynamics in that network. Not all nodes are equal, with some being more important than others.

Mark Granovetter is an influential American sociologist at Stanford University who pretty much invented the field of social network theory. He was talking about the dynamics of the spread of information in social networks—in a paper called "The Strength of Weak Ties"—way back in 1973. That was 11 years before Mark Zuckerberg was born.[12]

Granovetter argued that the strength of social relationships in social networks is a function of their duration, emotional intensity, intimacy, and exchange of services. His ideas are visualized every day on Facebook, Twitter, Instagram, and Snapchat. But they are also manifested inside every market where we see network dynamics taking over, and inside every organization where people interact and exchange information.

But what we are now observing is that most of these networks demonstrate a network behavior called the *rich club* phenomenon. The name arises from the analogy with social systems (often country clubs, alumni networks, or elite social circles), where highly central individuals—those who are rich in connections—often form a highly interconnected club. These rich clubs are often extraordinarily influential and extremely powerful, and are the engines of growth, progress, and wealth.

This rich club phenomenon in networks occurs when the hubs of a network tend to be more densely connected among themselves than nodes of a lower degree. We see the phenomenon everywhere now. For example, in power grids, some strongly connected power stations can easily distribute the power load of one station to another to reduce the possibility of critical failures. We see rich clubs in flight patterns and logistics networks. We see rich clubs inside the neurons in our brains. Some power stations are more important than others, some airports are more important than others, and some neurons are more important than others.

Even in our simple day-to-day activities, the creation of rich club networks is crucial. A study recently found that a group of college students who started interacting very intensely during the first week of class and continued to exchange information throughout the semester—predominantly with one another—performed much better than others in the class.

These students had, probably without realizing it, formed a rich club and reaped its benefits for their academic journey.

As Pareto would say, there is an elite group of connections inside networks that are often real engines of innovation. Even our human brain is now shown to act as a network, and there's a rich club inside our gray matter. The neural rich club in our cranium makes up only a small piece of the overall wiring of our brain, but it carries a massive amount of information flow.

LEVERAGE THE POWER OF NETWORKS

My fundamental belief is that organizations have to become networks if they want to survive. The essence of this belief is that companies now live in a world where the outside markets have become networks of information. If the outside is a network, the inside has to become a network as well.

Anyone can leverage the power of networks, but in order to do so, you need to understand the laws that govern them. I think networks are the blueprint for tomorrow's organizations, based on three fundamental axioms:

1. Information flows faster through a network.
2. Intelligence filters faster through a network.
3. Innovation flows faster through a network.

Allow me to expand on these axioms.

1. Information Flows Faster Through a Network

Today we are truly living in an information revolution. Access to information is ubiquitous, flowing faster than ever before. But information also becomes stale faster than ever before.

The essence of big data is *not* volume. Big data is not just about large data sets of information. Big data is about the velocity of information. The biggest shift that we see happening is that information used to be contained, but now information has to flow.

The great advantage of networks is that information flows faster in a network than it does in a hierarchy. Essentially, networks naturally turn information from ponds into rivers. Today's organizations need these rivers of information to handle the speed needs of the network society.

Probably the best-known example of this is gossip. Everyone knows that gossip travels through an organization faster than any official information travels through a company's hierarchy. Gossip favors the rich club of highly connected people who are in the know. But if you leverage the rich clubs of innovation and customer flow, you can move faster with your organization than ever before. Internet memes are another example. As silly as they are, they run like wildfire through our social networks. There are few people who are not familiar with grumpy cat or the "Charlie bit my finger" clip. These gems of "strategic" information would never have reached us if it hadn't been for networks.

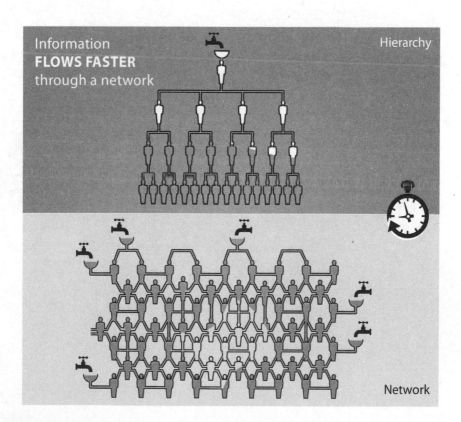

2.   Intelligence Filters Faster Through a Network

The second axiom deals with intelligence. Information, of course, is only half the picture. It's the insights, patterns, and intelligence that really matter. As Clay Shirky observed, most companies don't suffer from information overload; they suffer from filter failure. We just don't have the right information filters to distill the intelligence contained in the information.

This is not new. In 1982, John Naisbitt predicted that companies "would be overloaded with information, but starved of knowledge."[13]

But what we now observe in outside networks is that the network is an extremely effective filter. Feedback on social networks, ratings of participants, and comments on topics provide extraordinarily powerful filter mechanisms.

Companies that can harness the use of networks as filters will be able to distill intelligence from information much faster. Rich clubs unleash powerful filter mechanisms. We must discover them and understand the patterns and filters that turn information flow into practical filter mechanisms.

3.   Innovation Flows Faster Through a Network

The world has become innovation-obsessed. The clock speed of innovation is accelerating, and the chase for the *next new thing* is exhilarating. But many companies can't cope. Many organizations feel that their essential inertia is dragging them down, and that they can't compete in this increasingly fast race.

Many companies have extremely intelligent people in their organizations, but they're locked in old-style hierarchical command-and-control structures. This is what keeps innovation from flowing and holds many organizations back.

The third axiom states that innovation flows faster in a network than in a hierarchy. Start-ups are building on the work of others; open-source initiatives build on the intelligence of the collective; innovation feeds on the multiplicity and diversity of the network.

The U.S. Defense Advanced Research Projects Agency (DARPA) usually takes about five years to develop, build, and test a new vehicle. Because it

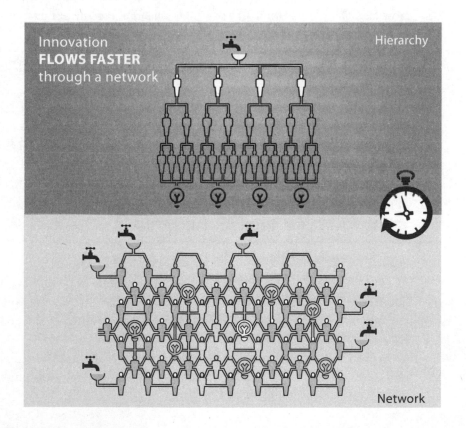

wanted to innovate faster, it collaborated with a small auto manufacturer, Local Motors. The latter often turns to crowdsourcing—a very powerful way of leveraging the network through open innovation—when it comes to product development. It launched a $10,000 prize competition to design the vehicle. In less than five months, it had a fully tested and operational vehicle ready. Talk about speeding up innovation: from five years to five months.[14]

Companies that can mobilize the power of the network internally will be able to let innovation flow much faster than in the traditional corporate hierarchy.

These three axioms form the basis for how companies should think about becoming more networklike, unleashing the power of the graphs, and leveraging the potential of hidden rich clubs. Information, intelligence, and innovation are the lifeblood of network-age organizations, and that's why you should understand how to foster them in your network.

Crowdfunding platforms like Kickstarter are a beautiful example of how networks can filter the best ideas and innovations. What better way is there to find out whether something will catch on than to find out whether the crowd is willing to pay for it—to see if it is willing to literally put its money where its mouth is?

An increasing number of companies are therefore using these kinds of platforms not just for fundraising, but to uncover market insights fast and efficiently. Tesla Motors, one of the most innovative companies in the world, has done just that, but even smarter: it requested advance reservation fees from customers of $5,000 per car. It not only succeeded in knowing exactly what the demand would be, but also took care of its own working capital at zero percent interest, rather than having to pay a bank 10 to 15 percent.[15]

THE TOPOLOGY OF YOUR ORGANIZATION

If you want to understand how to prosper in the age of networks, you will have to understand the topology of your connections. You will have to figure out the topology of how your customers behave in their networks, and understand how the topology of your internal networks of skills and information behave.

You will have to figure out what the rich clubs in the outside customer networks are, and you will have to get in touch with the rich clubs inside your own organization's network.

Claude Shannon laid the foundation for information theory, but its focus was on the transmission of information. It did not make a judgment about the meaning or relevance of that information. It did not interpret the information; it merely ensured that it was transferred and transmitted correctly, and interpreted in the right way at the other end.

Today, we have to take things up a notch. We have to make sense of the information flow in both outside and inside networks. We have to understand the meaning contained in the patterns and graphs that make up our world today.

Companies that understand how to interpret the flow of their customers and their employees will grow and thrive. Those that use the power of data science and big data to understand the predictive possibilities of interpreting the flow will be the winners in the age of networks.

May the flow be with you.

EDUCATION
IN THE AGE OF NETWORKS

The education model that we have in place today in most parts of the civilized world has its roots in the schools of the Kingdom of Prussia in the early eighteenth century.

King Frederick William I of Prussia should be the patron saint of the current educational system. He was a remarkable tyrant, concerning himself with almost every little detail in the running of his kingdom. The godfather of all micromanagers, he personally dictated the *Manual of Regulations for State Officials*, which contained 35 chapters precisely detailing the duties of every public servant. The Soldier-King, as he was known, never started a war, but he loved military-style discipline. When he died in 1740, his rigid control of the state and the economy ensured that Prussia was fabulously wealthy.

But the true legacy of King Frederick is the bell that rings after every class period in school. The division of the day into one hour per subject, the assignment of teachers to groups of students in class, and even the idea of homework all came from his rulebook.

In 1717, King Frederick decreed the compulsory education of children between the ages of five and twelve. The Prussian educational system was designed to turn savage children into disciplined citizens, and to take education out of the hands of the family and the church with five key objectives in mind:

- Obedient workers for the mines
- Obedient soldiers for the army
- Well-subordinated civil servants for government
- Well-subordinated clerks for industry
- Citizens who thought alike about major issues

Every individual had to become convinced, in the core of his being, that the king's decisions were always right. The system turned out to be the most effective way, by far, to instill social obedience in the country's citizens. Soon, the system spread across the world. Napoleon used it to control the education of the masses for his empire. Before long it spread to the new continent as well. Educator John Taylor Gatto describes how the system came to America:[1]

> A small number of passionate ideological leaders visited Prussia in the first half of the 19th Century, fell in love with the order, obedience, and efficiency of its educational system and campaigned relentlessly thereafter to bring the Prussian vision to our shores. To do that, children would have to be removed from their parents and inappropriate cultural influences.

Horace Mann is seen by historians as the father of the common school movement in the United States. He was the brother-in-law of the author Nathaniel Hawthorne, and he served in the Massachusetts House of Representatives. Later he was elected to the U.S. House of Representatives. In 1843, Mann traveled to Germany to investigate how the educational system worked. When he returned, he lobbied to have the Prussian model become the standard for American education.

Mann succeeded in his mission, and his aims at the time were extremely noble. He hoped that by bringing all children of all classes together, the system could impart a common learning experience. It would create the opportunity for the less fortunate to advance, and would, in effect, "equalize the conditions of men." Mann argued that Prussian-style schools would help students who did not receive appropriate discipline at home: "Instilling values such as obedience to authority, promptness in attendance, and organizing the time according to bell ringing helps students prepare for future employment."

Well, for future employment in the nineteenth century, that is.

The problem is that the world has moved on at breakneck speed. What worries me as a parent is that the curriculum employed in the schools is still pretty much the same as the one I experienced 30 years ago. The world has seen some of the biggest changes in history come about during that period, but the subjects of study remain largely the same. What worries me even more is that the system of classroom education divided into chunks of subject matter is the same as it was in the glory days of the Soldier-King.

THE ARTIFICIAL INTELLIGENCE ROCK STAR

Sebastian Burkhard Thrun was born in 1967 in Solingen, Germany. Located in the heart of the Rhineland, Solingen is known primarily for the manufacture of fine swords, knives, scissors, and razors. Thrun certainly has a brain as sharp as a knife, and when he attended university, he was quickly noticed as a mathematics wunderkind. As part of his thesis project, Thrun created a robot called Rhino that could give guided tours of the university and a local museum. He got a PhD (summa cum laude) in 1995 in computer science and statistics, and subsequently moved to Carnegie Mellon University to become codirector of the Robot Learning Laboratory.

Thrun was obsessed with understanding how to make machines think. He desperately wanted to turn the dodgy old field of artificial intelligence into reality, and to build a robot that would truly come alive. In 2001, Thrun spent a sabbatical year at Stanford University; he joined the faculty full-time in 2003, when he became part of the Stanford Artificial Intelligence Laboratory. Thrun soon became a Google Fellow. He worked closely on the development of Google Street View as well as the Google driverless car system.

Building a robot that could think was his initial plan, but building a car that could drive itself came pretty close to being incredibly awesome. Thrun became a cult professor, packing a 200-seat auditorium to the brim with enthralled students.

And then, in 2011, he became an international phenomenon when he launched the first global MOOC, or massively open online course. The idea came from the frustration Thrun felt about the limited audience he could fit into his auditorium: "What are 200 students in an age when billions of people around the world are connected to the Internet?" So he thought, why don't we move the whole thing online? Instead of limiting his class to Stanford students, who forked over more than $50,000 for the privilege, why don't we open up the course to the world? He decided to create an online course, open to everyone, together with Peter Norvig, the head of research at Google.

The results were mind-boggling. More than 160,000 students from around the world—every country in the world except North Korea—applied for the online artificial intelligence course. Students from the ages of 10 to 70 years followed his courses online, did homework assignments, and eventually took the exams—online. Out of the 160,000 students who enrolled, 23,000 eventually obtained a certificate of completion.

The first MOOC was born, and it changed the history of education forever. Thrun was a son of the remnants of the Prussian Empire, and he was certainly educated in the German school system, which still closely follows the model that King Frederick conceived. But Thrun's online experiment changed that model.

In 2012, Sebastian Thrun announced that he was leaving Stanford to focus on Udacity (www.udacity.com), an educational platform designed to commercialize the experience of his artificial intelligence course. Funded with $300,000 of Thrun's personal money, the new start-up quickly got more than $15 million in venture capital funding. Udacity aims to educate the world in science, technology, engineering, entrepreneurship, and mathematics—all online. One of its programs, in collaboration with the Georgia Institute of Technology, will offer a full computer science master's degree program for only $6,600, a fraction of what it would cost in a conventional university setting.

TOWARD A NETWORK OF LEARNING

As you would imagine, these developments are shaking up educational institutions. It makes one wonder how universities will remain relevant when the Udacities of this world become a valid way to get a degree or learn a new skill. Will universities even survive in the age of networks?

In a brilliant article published in *Time* magazine in 2013, Leo Reif, the president of MIT, argues that online learning could reinvent higher education, and that digital learning is probably the most important innovation in the world of education since the printing press was invented.[2]

Leo Reif was born in Venezuela and received his doctorate in electrical engineering from Stanford in 1979. He joined the MIT faculty in 1980, and eventually went on to become the head of MIT's Department of Computer Science. In 2012, Reif was appointed the seventeenth president of MIT.

Reif understands what technology can do to transform education in the age of networks: "When the class of 2025 arrives on campus, technology will have reshaped the entire concept of college in ways we cannot yet predict. Those transformations may change the whole equation, from access to effectiveness to cost."

In 2002, MIT started an initiative called OpenCourseWare that has posted virtually all of the university's course materials online, for free. It has attracted more than 150 million learners worldwide. MIT also launched an online learning platform called edX, together with its archrival Harvard, that has enrolled more than 1.25 million unique learners in just 17 months. As Reif points out: "That's more than 10 times the number of living MIT graduates."

But can an online platform have the same impact as a real teacher can? Can an online educational network give the same quality of education as a classroom experience? According to Daphne Koller, the founder of Coursera, it can be even better.

Daphne is an impressive woman. She was born in Israel in 1968, and received a bachelor's degree from the Hebrew University of Jerusalem in 1985 when she was only 17. Like Sebastian Thrun, she became fascinated with the field of artificial intelligence and gravitated toward Stanford. Together with the head of the Stanford Artificial Intelligence Lab, Andrew Ng, Daphne founded Coursera in 2012.

Coursera is an educational technology company offering massive, open online courses in subjects ranging from engineering to biology, from computer science to the humanities. It has received more than $65 million in venture capital, and today it offers hundreds of courses online for anyone to take, and for free. Its most popular online courses attract up to 200,000 students.

These large numbers, combined with network effects, are changing learning. But are they making it better?

Daphne Koller thinks so. She argues that online education based on the concept of mastery learning is much better than any classroom lecture experience can be. "In traditional learning," Koller says, "if a student does her homework and doesn't do well, she simply gets a low score on the assignment, and instruction moves to the next topic, providing the student a poor basis for learning the next concept. On our platform, we can give immediate feedback on that concept the student did not understand. This can increase student performance by about one standard deviation over more traditional forms of instruction."

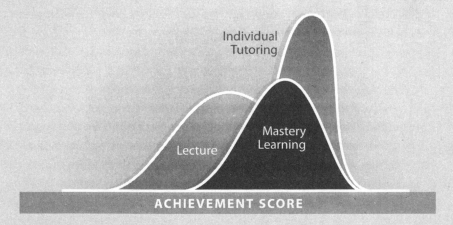

Nothing will ever beat individual tutoring. A one-on-one learning experience from master to student is the best possible form of education. Nothing can replace a close personal connection with an inspiring mentor. But the mastery learning concept applied in the field of online education can be much better than traditional lecture-based teaching, and can come pretty darn close to individual tutoring.

The age of networks also allows us to use the power of networking in the student population. Platforms like Coursera are using peer assessments, in which students evaluate and even grade the work of their fellow students. This method has been shown to result in extremely accurate feedback. It also provides a valuable learning experience for the students who are doing the grading.[3]

The age of networks won't wipe out the Stanfords or MITs of the world—at least, not anytime soon. But as Leo Reif says, "It opens possibilities for billions of human beings who have little or no other access to higher learning."

CAN YOU TEACH ME IN NINE MINUTES?

The higher education system is in full transition to the era of networks. Universities will have to reinvent themselves, and they will have to understand how the possibilities of networked learning can augment, complement, and enhance their offerings.

But the fundamental challenge for our education system is understanding how to use the power of networks for children before they reach college age.

Today's schoolchildren have grown up on YouTube and Google, have learned the art of networking with Facebook and Snapchat, and are bored out of their skulls in their Prussian-style classrooms.

The biggest mistake many educators and schools have made is to think that in the world of the New Normal, all they had to do was throw technology at the problem. That, in many cases, has only made things worse.

The most famous case occurred in 2013 in the Los Angeles Unified School District, which wanted to equip more than 30,000 students in 47 schools from kindergarten through twelfth grade with iPads, an investment of more than $50 million. The iPads had educational software preinstalled, and were locked down so that students couldn't access social networks or YouTube to distract them in class. In this day and age, that kind of security mechanism is to young teenagers what a red cloth is to a raging bull. Within just days of the the iPads being handed out, a couple of young hackers at Westchester High School had found a way to bypass the security software so that they could happily update their Facebook pages and stream music. The L.A. Unified School District promptly ordered that all tablets in the schools where the security system had been compromised be returned, and put the project on hold.

This is exactly what happens when we don't solve the fundamental issue. The failed experiment in Los Angeles proves that there is no point pouring expensive technology into classrooms that are still being run the old way. If we don't reinvent the education model for the age of networks, there is not enough technology in the world to make any difference in the outcome.

Let's not confuse the format of education with its mission. The mission of education is to prepare young people for their future lives, to help them innovate and prosper in the world they will be living in. Information has become a flow in the age of networks, so learning has to become a flow as well. It's not about iPads or Smart Boards in classrooms. It's not about technology.

We have to fundamentally flip the educational model by preparing courses, subjects, and curricula for those in the networked generation, who consume information in the ways they learned at home. Inspired by the disruptive MOOClike innovators, old-school universities have increasingly understood that they need to adapt their old models to fit the way their environment and their pupils have changed. Some of their biggest challenges—next to an in-depth change in culture that surmounts just "becoming digital"—will involve their tuition fees and status. If Harvard lowers its fees for online diplomas, its margins will shrink—and worse: then "everyone" can be an alumnus. Where's the prestige in that? They need to understand that a networked environment is flat, though. It is a meritocracy. Status and prestige are worthless there. So, yes, some Ivy League schools will definitely need to do some adapting.[4]

My first TEDx talk was one of the hardest things I've ever done. It was much more difficult to get my message across in the very strict 18-minute format than in the hourlong lectures I typically give on the lecture circuit. Just as Twitter, with a limit of 140 characters, forces you to be very disciplined and concise in what you write, TED forces you to be very clear in getting your message across. The latest generation of TED talks has now reduced that format to a mere nine minutes.

The original TED (Technology, Entertainment, and Design) conference was held way back in 1984. The event was conceived by Richard Wurman, a legendary architect, writer, and graphic designer. Wurman saw different fields converging and understood that people in technology could benefit from learning what was going on in design and entertainment. He wanted to create networks of influencers from different backgrounds. His first TED conference was a resounding success and a colossal financial failure. The original conference had demonstrations of the very first Apple Macintosh computer, which had just been introduced, and keynote presentations by people such as Nicholas Negroponte, then the head of MIT Media Lab; Benoit Mandelbrot, who had invented the mathematics of fractals; and Stewart Brand, the founder of the *Whole Earth Catalog*.

Richard Wurman understood very early on that information would reshape society. In fact, he coined the term *information architect* in 1976. As Wurman says, "I thought the explosion of data needed an architecture, needed a series of systems, needed systemic design and a series of performance criteria to measure it."

After the initial TED conference flopped financially, it took six years before the second conference was organized. From then on, however, the TED conference has been held annually in Long Beach, California. And in 2009 the TED organization started granting licenses to third parties to organize independent TEDlike events internationally, called TEDx. TED has now become a global phenomenon that has given the world access to a truly networked set of "ideas worth spreading."

The biggest impact of TED is that it has shown the world that you can educate, inspire, or entice people with content, ideas, and information in a format that is tailored to our current consumption patterns. Those patterns are simply not compatible with the Prussian system.

THE FLIP IN THE CLASSROOM

Salman Khan has become the poster boy for the revolution in the educational system—the beacon of hope for those who believe that it is time to flip the classroom.

Today, the Khan Academy is the most prominent online education platform that offers free lessons in a wide spectrum of subjects that are relevant to high school students and teachers.

Khan is an unlikely hero. He was born and raised in New Orleans. When he left the Bayou State to study at MIT, he promised his cousin, Nadia, that he would keep tutoring her in mathematics, as he had done in the past. He promised her that he would use the Internet as a tool. First, he used

Yahoo!'s Doodle notepad, but when other relatives and friends asked him to tutor them as well, he decided that it would be more practical to put it on YouTube. His videos went viral, receiving more than 300 million views in just a few years. Students from around the world were attracted to Khan's concise, practical, and relaxed teaching method, which gave him the idea of starting the Khan Academy. Today, his concept has produced more than 4,300 video lessons and inspired millions of students. His mission is to provide "a free world-class education for anyone anywhere." In 2012, *Forbes* magazine put Salman on its cover; the cover story was about how to flip the $1 trillion education market.

The Scottish philosopher David Hume once famously said, "It is not reason which is the guide of life, but custom." Too often, we extrapolate the past, shave off a few rough edges, and keep doing what we've always done. Unfortunately, the educational system has become a glaring example.

King Frederick's educational system was shaped by a vision of work, productivity, and adulthood that, arguably, remained relevant throughout the Industrial Age. But the age of networks that we're now entering demands a totally fresh perspective on how we will work, be productive, and live our adult lives.

We won't solve the problem by throwing technology at the classroom. We have to flip the model. We have to reboot the education system to embrace the age of networks.[5]

We owe it to our kids.

CHAPTER 5

WHEN MARKETS STOP BEING MARKETS

The ability to dictate the conversation with the consumer is fading quickly. Markets have transformed themselves into networks of intelligence, which follow different rules and observe different behavioral patterns.

Marketers have long relied on technology to help them reach consumers more efficiently and exert control over markets. But now technology is changing the rules for how those customers can be reached—and making control impossible. To survive, marketers must learn how to influence information networks, a process that begins with earning trust.

Marketing has always had a curious and intertwined relationship with technology. Whenever there has been an innovative surge of technology, marketing has rushed to the scene to see how that surge could be exploited to reach a bigger audience, faster, and with more impact.

Most often, technology has allowed people to spread the word faster and faster.

The invention of the printing press in the West is attributed to the German printer Johannes Gutenberg in 1450, although the first movable-type printing technology had already been developed in China by the Han printer Bi Sheng at the start of the eleventh century. A goldsmith by profession, Gutenberg developed his printing system by both adapting existing technologies and making inventions of his own. The results were spectacular: a single Renaissance printing press could produce more than 3,000 pages per workday. Soon, books of bestselling authors such as Luther and Erasmus were being sold by the thousands in their lifetimes.

In Renaissance Europe, the arrival of the printing press introduced the era of mass communication, which altered the fabric of society forever. Technology enabled a revolution in communication, which in turn produced a revolution in the way we live our lives.

This information revolution transcended borders, broke the monopoly of the literary elite in education and learning, and gave birth to an educated and empowered middle class. Of course, not all of that was due to the invention of the printing press. The emerging spirit of capitalism and the eagerness of the middle class to rise through knowledge and learning laid the foundation for the technology of the printing press to spread so rapidly. It changed the concept of books and authors forever. For many works prior to the printing press, the name of the author has been entirely lost. The advent of the printing press made the concept of authorship more meaningful—and immensely more profitable. It suddenly was important who had said or written what, and what the precise formulation (and timing) of information was.

This helped fuel the scientific revolution, in which authors (scientists) could publish and communicate their discoveries in widely disseminated journals. Today, when we pick up a book or a magazine, we rarely remember that print was once a new medium—one that started a societal metamorphosis.

BOOK BIND

Although we have immortalized Gutenberg, he himself did not fare very well. Suffering from the ancient curse of the inventor, he was, in effect, a failure as a technology start-up. Conceiving and building a printing press back in 1450 wasn't so different from getting a start-up off the ground in a Palo Alto garage. In both cases, the fundamental problem is funding. If you don't have the cash, your ideas are worthless.

Gutenberg had already taken out a substantial loan from his brother-in-law Arnold Gelthus back in 1448, which had allowed him to build a prototype printing press and collaborate with the Master of the Playing Cards, a Swiss master engraver whose identity has never been traced, but who produced amazingly beautiful copper engravings. Two years later, in 1450, Gutenberg's printing press was in operation. However, the money had run out. The loan from his brother-in-law (the friends-family-fools fund-raising) had been spent. So Gutenberg had to look for other rounds of funding.

He convinced wealthy moneylender Johann Fust to lend him 1,600 guilders. As part of the deal, Peter Schöffer, Fust's son-in-law, joined the enterprise. Schöffer had worked as a scribe in Paris and is believed to have designed some of the first typefaces, so he was knowledgeable about the subject. But it is never a good idea for a start-up to have a venture capitalist too closely involved in its day-to-day operations.

The business started picking up, and Gutenberg and his team began printing lucrative projects such as Latin grammars, but the breakthrough came in 1455 with the Gutenberg Bible. Only about 180 copies were printed, most of them on paper, but some on vellum. They were valuable at the time, but today they are priceless. Only 48 copies, or substantial portions, survive today. They are considered to be the most valuable books in the world, even though a complete set has not been sold since 1978.

Although the Bible project was a success, Gutenberg got into financial trouble—big time. By 1456, a massive dispute had arisen between Gutenberg and Fust, who demanded his money back, accusing Gutenberg of misusing the funds. Meanwhile, the expenses of the Bible project had escalated, and Gutenberg's debt now exceeded 20,000 guilders.

Fust brought a suit against Gutenberg to recover the money he had lent, and the court decided in favor of Fust, giving him control over the Bible-printing workshop and half of all printed Bibles. Gutenberg was effectively bankrupt, and he never recovered his glory or fame. He died broken and embittered in 1468 in the city of Mainz.

There are two valuable lessons are to be learned from the Gutenberg story. The first is that the inventor of a breakthrough technology rarely reaps the full benefits of his creations. In fact, being a close follower of a true innovator might be a better strategy for success—just ask Bill Gates. The second big lesson here is that when you get a greedy venture capitalist who is too close to the action and too involved in your business, you're sleeping with the enemy. So if you ever launch an entrepreneurial start-up and your financier says that his amazingly brilliant son-in-law would be perfect as your COO, think of poor Gutenberg.

The printing press turned out to be a long-lived technology. It was John Lienhard who introduced us to the notion of short-lived technologies: bridging technologies that flare up violently and are fiercely relevant for a very brief period, then give rise to a more fundamental new paradigm.[1]

THE ORIGINAL SPEEDBOATS

One of the most beautiful of the short-lived technologies was the clipper ship. Clippers were extremely fast sailing ships that had three or more masts and a square rig. They were extremely narrow for their length (which gave them their sexy shape) and had huge total sail areas, but they could carry extremely limited amounts of bulk freight. They were the Formula One racing ships of the oceans, sailing primarily on the trade routes between the United Kingdom and its colonies in the east, and the New York–to–San Francisco route around Cape Horn during the California Gold Rush. Dutch clippers were built beginning in the 1850s for the tea trade and passenger service to Java.

Ocean shipping requires a trade-off between speed and capacity. Traditionally, the longstanding compromise was that cargo moved over the seas in slow-moving, high-capacity merchant vessels. But that balance was upset in 1845, when San Francisco became the golden gate to the western United States. The same booming economy that drove shipping to California also drove the market for Chinese tea. In a very short time, shipping rates rose from $10 per ton of cargo to over $60 per ton. And as shipping rates exploded, it became profitable to build and operate ships that functioned more like racing vessels than like traditional slow-moving cargo carriers. Enter the clippers.

Masts rose into the sky; hulls developed a knife-edged bow. Speed was more important than carrying capacity. It would be the equivalent of equipping the UPS guys with Maseratis instead of the traditional brown vans. But today, we can really see clippers only in pictures. Only a few hundred were ever built. The financial boom ended in 1855, and after that, the ships vanished. The new steamships took over, and the economics of the ocean voyage again gave way to the slow-moving but massive load-carrying vessels, ocean freight carriers that would become bigger and bigger, and uglier and uglier. Speed was no longer worth the load-carrying inefficiency of these glorious sailing vessels.

In the last hundred years, in the field of marketing, we have seen myriads of new technologies that have aided us in reaching out to consumers, delivering the right message to the right audience, and trying to convince people to buy our products or services. In the last 10 years, we've seen an almost Cambrian explosion of new marketing technologies. The question will be: Which new advances in communication will be clipper ships, and which will prove to be truly long-lived technologies?

THE (OLD) HEART OF MARKETING

The science of marketing was invented to help sales. At least, that's the story that the giants of marketing like to tell. Marketing was created to generate insight into what consumers crave so that companies could produce exactly what the market wanted.

Philip Kotler is considered by many to be the father of modern marketing. Kotler got his PhD in economics at Massachusetts Institute of Technology in 1956. He then moved into marketing, which he regarded as an essential part of economics, and started teaching the subject in 1962 at the Kellogg School of Management at Northwestern University.

Kotler's fundamental belief was that marketing is not a separate thing. Instead, he believed that the marketing purpose of meeting the needs of consumers had to be put at the heart of every company strategy and be practised by all managers. According to Kotler, "the organization's marketing task is to determine the needs, wants and interests of target markets and to achieve the desired results more effectively and efficiently than competitors, in a way that preserves or enhances the consumer's or society's well-being." He's also the man behind the legendary but slightly outdated *Four Ps*: product, price, place. and promotion.[2]

If you look at the evolution of the science of marketing, in the beginning, marketing was really about relaying a message from the producer to the consumer. And it was often a very narrow, one-way conversation.

When a company had a new product or service offering, it wanted to tell the market about that offering. And that's exactly what marketing did. The clever marketers of a company would hire the crafty Don Drapers of the world to shape a cunning message that they could hurl onto the radios and television screens of consumers.

Again, the intertwined relationship between technology and marketing became a testing ground for constantly finding new ways to reach target audiences. Probably the finest example of that was the birth of the soap opera. The pioneering technology of radio was used in an effort to reach a predominantly female audience with daily serialized dramas.

CONSUMERS TAKE OVER

But gradually, consumers started getting vocal. They wanted to be heard. They wanted to be involved. It was not enough for companies to deliver messages to the consumer; they needed to build a conversation with the consumer.

The advent of the New Normal accelerated this process. Marketers quickly seized the opportunities of social media, becoming *conversation managers* who needed to understand how to converse with their consumer base. It was not about communicating a message any longer; it was about building strategic conversations, about learning the art of social conversation.

Today we are at a point where many markets are about to flip—that is, turn into networks of intelligence. Consumers have become extremely informed networked thinkers who are influenced by what they hear, see, and read on the network of intelligence that surrounds them, including peers who are always within reach. They trust one another more than they trust some commercial message on the TV or radio. Consumers have found one another. There's no way that this Pandora's box will close again.

This means that the old marketing concept of *funnels* is dead in the water.

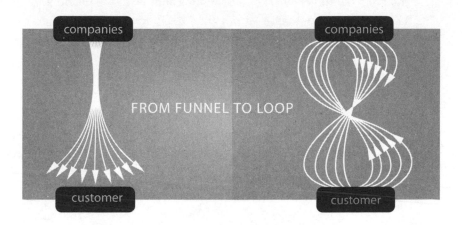

GOOD CLEAN FUNNEL

Funnels have been used in marketing forever. The idea is simple: there is a population of potential users of your product or service. This is your universe. Your job as a marketer is to take this universe through the funnel. In other words, to narrow that population down to the really qualified leads—people who, given the right information, will definitely want to buy—and to guide those leads to that final decision point, or *moment of truth*. That's the moment when they hopefully choose your product instead of the other brand, or pick up the phone and order your service instead of the alternative. The funnel is a powerful metaphor. But it doesn't work once your market has flipped.

Consumers have become intelligent, informed, and empowered. They trust the network more than they trust you.

Before making a purchase, an informed consumer will gather all the information necessary. Of course TV and radio are in the mix, but more and more consumers will be seeking out websites, blogs, experiences on Facebook, Twitter, or anything else. And when they get to the moment of truth, they will be influenced more by their own network of intelligence than by your marketing efforts.

According to research by Google, in 2013 almost half of the population used their mobile phones in a store when shopping, to find out more about a product. More important, almost 20 percent of shoppers changed their mind in the store based on the information they get from their mobile phones.[3]

So, the moment of truth isn't what it used to be. And even after that moment, consumers will keep communicating with the network. They will talk about their purchase, about their decisions, and about their experiences. In some cases, consumers will actually spend more time doing this after the moment of truth than before it. This means that the old funnel has become a loop.

But it also means that the power balance between producer and consumer has fundamentally changed.

Markets have now become networks of intelligence that are in constant flux. You can't control a market anymore. You actually have to work very hard just to follow a market and observe its flow.

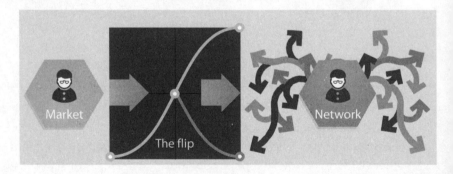

Talk about a flip. The new paradigm will have consumers who are more informed than you are, in markets that have become networks that can change and adapt faster than you can. It's a whole new game.

Marketing has gone from a very linear process, where the marketer sent out a message to the customer, to a very complex process, where different sources are influencing a buying decision, and where the consumer has also turned into a source of information.

A REAL-LIFE *MAD MAN*

So what does that mean for advertising? It's too bad David Ogilvy isn't around to give his perspective on the matter. David Mackenzie Ogilvy, born in 1911, is widely hailed as the father of advertising. In 1962, *Time* magazine called him "the most sought-after wizard" in advertising.

It wasn't always clear that Ogilvy would become so influential. After failing his studies at Oxford, he moved to Paris, where he became an apprentice chef in the Majestic Hotel, one of the finest hotels in the city. After a year, he returned to Scotland and started selling AGA cooking stoves door-to-door, which he did with incredible success. He was so effective in his sales approach that AGA asked him to write a sales manual, *The Theory and Practice of Selling the AGA Cooker*. Impressed by this work, his older brother showed the manual to the management of a London-based advertising agency called Mather & Crowther, who offered David his first job in advertising.[4]

Ogilvy eventually wrote a classic book called *Confessions of an Advertising Man*, in which he told the story of how he came to understand the power of the advertising trade. One of my favorite stories in the book is this one: after just a few months in advertising, Ogilvy was assigned a client who wanted to advertise the opening of his hotel, but had a budget of only $500. Ogilvy realized that he couldn't use conventional advertising and media, so instead, he bought $500 worth of postcards. He sent invitations to everybody he found in the local telephone directory. The hotel opened with a full house. "I had tasted blood," concludes Ogilvy.[5]

In 1938, Ogilvy moved to the United States, where he went to work for George Gallup's Audience Research Institute in New Jersey. Ogilvy cites Gallup as one of the major influences on his thinking. Gallup emphasized meticulous research methods and the power of information and statistics. Advertising was moving to the world of economics, a powerful blend of statistics and appeal to human emotions.

After World War II, during which Ogilvy worked for the British Intelligence Service at the British embassy in Washington, he made a dramatic lifestyle switch and bought a farm in Lancaster County, Pennsylvania, where he lived among the Amish community. He loved the atmosphere of "serenity, abundance, and contentment," and he stayed in Pennsylvania for several years. Eventually he had to admit his limitations as a farmer and moved to Manhattan.

It was then that he built up his reputation, and the business that would allow him to make a lasting mark on the profession. Ogilvy started his own advertising agency, Ogilvy & Mather, with the backing of Mather & Crowther, the London agency then being run by his elder brother.

DISRUPTIVE FORCES

Ogilvy & Mather was built on the core principle of David Ogilvy's philosophy: the function of advertising is to sell, and successful advertising for any product depends on information about its consumer. Information was the new oil, even back in the sixties.

The agency was wildly successful, in no small part thanks to the spectacular charm and personality of its founder. In 1973, Ogilvy retired as chairman of Ogilvy & Mather and moved to Touffou, his utterly amazing estate in France, near Poitiers. In 1989, the Ogilvy Group was bought by WPP Group, a British company, for U.S. $864 million in a hostile takeover.

The legacy of David Ogilvy is that insight into the customer is at the heart of understanding how to influence the customer. My favorite quote is this one: "The consumer isn't a moron; she is your wife." He should know: he was married three times.

When we look at the episodes of *Mad Men* over different seasons, and watch Don Draper and his cronies spin stories for their clients, we slowly watch different technologies unfold. Print campaigns give way to radio commercials; radio yields to television advertisements. Technologies change, but there is no real change in the fundamental business model of marketing: understand your audience, segment it into addressable chunks, and find the right channel to get your message to each segment of the audience. In the age of the funnel, it worked wonderfully.

Now, not so much. The advent of the Internet in our daily lives unleashed one of the most disruptive forces since the invention of marketing itself: personalized, one-to-one customer interaction.

DON AND MARTHA

Push technologies were one of the biggest and most extremely short-lived flashes of the first Internet boom.

The idea behind push technology was simple: because computers are almost always going to be connected to the Internet, we can push interesting content to your computer. The more you use the system, the more we'll know about your interests and desires, and the more accurate the pushed content will be. Simple.

The pinnacle of the push-technology hype was a company called Point-Cast. It sprang to prominence in Silicon Valley in the late 1990s, when enthusiasm about the Internet was at its peak, but people were desperate for something more exciting than a mere website. PointCast came to the rescue, riding on the waves of push technology, but with a model that, in hindsight, was ridiculously flawed.

The idea behind PointCast was to push content to your computer, filling your screen with interesting and relevant content whenever your computer monitor went into sleep mode and the screen saver came on.

The hype went through the roof. Push technology was the hippest thing in the valley, and PointCast was the new Netscape. Venture capitalists from Sand Hill Road were piling up in the reception area.

Those VCs overlooked a now-obvious flaw: since screen savers come on only when you're not paying attention to your computer, there's a very good chance that you won't see the content. Perhaps they should have called it *PointLessCast*.

Nevertheless, in January 1997, News Corporation made an offer of $450 million to purchase the company outright. That was *after* the product was not performing as well as had been expected, partly because its traffic burdened corporate networks with excessive bandwidth use. News Corp's offer didn't go through, and things went downhill from there. A $250 million IPO was drawn up in 1998, but by then it was becoming clear that push technology might have been just hype, and was very short-lived. The company was sold in 1999 to Idealab for $7 million, a fraction of the capital that had flown into it. Idealab shut down the PointCast network the next year.

But the legacy of PointCast was not the concept of push technology. The real driver was the quest for personalized communication.

Don Peppers and Martha Rogers started talking about one-to-one marketing even before the World Wide Web came onto the scene in 1995. Don Peppers was the CEO of a direct-marketing company.[6]

Don and Martha's idea of customer-centricity was essentially the continuation of the Ogilvy philosophy: putting the customer first, and understanding everything you could about the customer. The difference was technology: the Internet gave marketers the essential technology required to deliver personalized messages.

Peppers and Rogers described personalized marketing as a four-phase process: identify potential customers; determine their needs and their lifetime value to the company; interact with customers to learn about them; and finally tailor the offering of products, services, and communications to the individual customers. There was no rocket science here, or groundbreaking concepts, but the underlying technology infrastructure had finally started to shape up to allow marketers to start making one-to-one dreams a reality.

One-to-one marketing suggested a version of the individualized service provided by the mom-and-pop store—except with millions of customers all over the globe. But the technology at the end of the twentieth century couldn't quite meet those expectations. Instead, one-to-one marketing became a scheme that really was all about mass customization—essentially the mass production of a shallow feeling of relevance.

As Peppers and Rogers themselves put it, "Mass customization, in marketing, manufacturing, call centers and management, is the use of flexible computer-aided systems to produce custom output. Those systems combine the low unit costs of mass production processes with the flexibility of individual customization." But customers wanted more than to be part of a "mass customization" of personalized experiences. They really wanted unique recommendations, tailored to their very own desires, and *truly* personal.

FROM FIREFLY TO AMAZON

The gold rush for personalization technology was on. In 1995, a group of MIT Media Lab engineers, led by the high-profile professor Pattie Maes, founded a company called Firefly. In essence, as *Wired* magazine described it, Firefly was a collaborative filter: a technology that asked users what they liked, learned their tastes, and then got them in touch with people who had similar tastes.

Back at MIT Media Lab, Maes, along with three grad students, had developed a program called Helpful Online Music Recommendations (HOMR). It used intelligent-agent technology to provide music lovers with suggestions about bands they might like.

Maes and company ran it as a website at first, changing the name to Firefly.com because it sounded better than HOMR. But they soon started to license its technology to others who wanted to provide collaborative filtering services for big names such as barnesandnoble.com and Yahoo!. Meanwhile, Firefly was gathering an immense amount of very sensitive information and profile data about its users, and was doing groundbreaking research on the frontiers of privacy.

In 1998, Microsoft purchased Firefly predominantly because of its innovations in privacy, but it shut down the website a year later.

Today we find it quite natural to go onto a website and be fed recommendations, but Firefly was the first to build this technology to scale. Today, probably the king of collaborative filtering is Amazon's recommendation engine. The traditional quip is, "Amazon knows what book you want before you know it yourself." And this isn't not far from the truth: when Amazon recommends a product on its site, it is clearly not a coincidence.

In essence, Amazon's recommendation system is based on a number of very simple elements: what a user has bought in the past, which items he has in his shopping cart, the items he's rated and liked, and what other customers have viewed and purchased.

In simple terms, the recommendation system is based on aggregating data about your browsing, purchasing, and reading habits, and then extrapolating these data to figure out what you would like to read next, based on what other customers with similar histories are reading. It turns out that we aren't as unique as we'd like to think.

RECOMMENDATIONS

Amazon doesn't disclose much about its underlying technology (Amazon doesn't disclose much, period). In a rare *Fortune* magazine interview about its recommendation system, a spokesperson from Amazon said: "Our mission is to delight our customers by allowing them to serendipitously discover great products."

And Amazon seems to be extremely good at that. How many times have we added something to our Amazon shopping cart, for just a couple of extra bucks, that we hadn't been searching for at all?

I believe the trick behind Amazon's strength as a recommendation engine is its unique combination of serendipity and trust. Amazon almost always shows consumers the products they are most likely to purchase, even if that recommendation makes the retailer less money than the alternative. Amazon has clearly realized that the more customers trust the recommendations, the more often they'll act on them, and the more money they'll spend with Amazon.

The publishers of books never knew their customers, the readers. They saw the bookstores as their customers. For a long time, publishers of books were the connection between authors and bookstores, and greatly neglected their eventual consumers, the reading public.

Today, Amazon is killing them. Not only has the distribution of physical books changed completely as a result of online retailing, but with the advent of e-books, the tables have turned completely. Amazon knows what books you buy, obviously, and with electronic books, it also knows what, where, and how quickly you read them. It sees which paragraphs you underline, highlight, and share. In essence, Amazon knows everything about your reading behavior. And book publishers know almost nothing.

The results? As author David Gaughran says, "The danger for large publishers is clear: they aren't just losing control of which books get published, but also which books get recommended."[7]

THE SECOND LIFE OF NETFLIX

Another extremely pivotal actor in the world of one-to-one is Netflix. Netflix today is the world's largest provider of on-demand streaming video over the Internet. When key episodes of top TV series first appear on Netflix, the service's traffic can constitute one-third of *all* Internet traffic in the United States. The company is a marvel of high-tech and brilliant online strategy, but it had very humble beginnings.

Also, quite silly beginnings, if you really think about it. Netflix got started as a service by *mailing* you DVDs via the postal service. That's right, in an envelope.

The founder of Netflix was Reed Hastings, born in Boston in 1960. When he finished high school, he spent a year selling Rainbow vacuum cleaners door-to-door. After getting a degree in mathematics, Hastings joined the Peace Corps in 1981, and went to teach high school math in Swaziland from 1983 to 1985. After returning from the Peace Corps, Hastings attended Stanford University and got a master's degree in computer science.

Hastings worked at a number of successful software companies in Silicon Valley before he cofounded Netflix with Marc Randolph in 1998, offering flat-rate rental-by-mail DVDs to customers in the United States.

He claims he got the idea for the DVD-by-mail service after getting a late fee for $40 for not having returned the VHS copy of *Apollo 13* to his local rental store on time. He then started to investigate the idea of creating a movie-rental business by mail. As he says: "I didn't know about DVDs, and then a friend of mine told me they were coming. I ran out to Tower Records in Santa Cruz, California, and mailed CDs to myself, just a disc in an envelope. It was a long 24 hours until the mail arrived back at my house, and I ripped them open and they were all in great shape. That was the big excitement point."

Netflix grew like wildfire, especially since there were plenty of people in the United States who lacked access to a full-sized movie-rental store nearby. But the growth of the Internet—and the eventual possibility of streaming videos over the network instead of mailing them—gave Netflix a totally new perspective on its business.

As a matter of fact, Netflix is one of those legendary companies that managed to turn itself around during a period of technological turmoil and seize the opportunity of the new. Netflix became a huge company in the business of mailing DVDs to customers—a short-lived technology if there ever was one. By seizing the opportunity to stream videos online, Netflix avoided the fate of slower-moving companies.

In the last couple of years—since its 2011 Qwikster stumble, which is another story[8]—Netflix has seen impressive growth in customers, offerings, and geographical reach. It's now seen as a major player in the future of television over the Internet.

And the key to unlocking that future is personalization.

GETTING PERSONAL

Just as Amazon recommends books that you might like to read, Netflix offers predictions of movies that a user might like to watch based on the user's previous ratings and watching habits. It also takes into account the characteristics of each film, such as genre and actors. It's Firefly all over again.

But Netflix actually took it one step further. It leveraged the power of open innovation and network intelligence, and set up a contest, the Netflix Challenge, that offered a grand prize of $1,000,000 in 2006 to the team of engineers that could take a data set of 100 million movie ratings and develop an algorithm that was more than 10 percent better than the existing Netflix recommendation system. In other words, if you could build a better one-to-one recommendation system than Netflix, you would win a million bucks.

Netflix sponsored this competition from 2006 to 2009, energizing the search for new and more accurate recommendation algorithms. On September 21, 2009, the grand prize of U.S. $1,000,000 was awarded to BellKor's Pragmatic Chaos team. A second contest was planned, but it was ultimately canceled in response to an ongoing lawsuit and concerns regarding privacy from the Federal Trade Commission.

It should be made clear that this research into personalization is pushing the boundaries of privacy almost daily. We all know the creepy example of Target sending a teenage girl coupons for baby products, having learned that she was pregnant through big data analytics—even before her father learned it. But it is also clear that in order to survive in a world of information overload, in a world of seemingly endless possibilities, the algorithms are winning. In 2013, Amazon Instant Video received a Technology & Engineering Emmy award for its recommendation engine.

As Amazon's Bill Carr says, "Our goal is to give customers the best possible movie and TV watching experience. That means both enabling customers to find exactly what they're looking for and helping them discover new TV shows and movies in a personalized way."

But what do people really want? What really drives a human's behavior, and how well can we tailor products to your individual tastes and needs? How do we get inside your brain to figure out how we can market to you?

Not with a screen saver, that's for sure.

In the last couple of years, we have not only seen an incredible advance in the digital technologies that allow us to understand consumer behavior and influence you with content. We've also seen an explosion of insight into how our brain actually works, and why we react the way we do.

The research is both surprising and potentially disturbing. It turns out that we have much less conscious control over our actions and decisions than we originally believed.

THE POPCORN EXPERIMENT

The controversy over the role of the unconscious in marketing has been a topic of debate ever since the infamous research into subliminal advertising by James McDonald Vicary in 1957. Vicary was a market researcher, doing work on impulse buying and word association, when he conducted an experiment in a movie theater in Fort Lee, a suburb of New York City, during the warm summer of 1957. For six weeks, he injected extremely brief subliminal messages into a screening of *Picnic*, starring William Holden and Kim Novak. The messages, urging the moviegoers to "eat popcorn" or "drink Coca-Cola," appeared for only 0.03 second. Viewers of the movie could not consciously see the messages; the idea was to see whether regular advertisements, which the viewers often found boring, could be eliminated.

The results were spectacular. Sales of Coca-Cola jumped 18 percent and sales of popcorn more than 56 percent. When the trade magazine *Advertising Age* got hold of the story, it created widespread panic. The public went wild about the possibilities of mind control, thought poisoning and brainwashing. It even led to a widespread investigation by none other than the CIA, which, after a lengthy study, decided to move for a ban on subliminal advertising, stating, "Certain individuals can at certain times and under certain circumstances be influenced to act abnormally without awareness of the influence."

Unfortunately, the whole experiment was false. The results could never be reproduced in any other setting, and in 1962 Vicary admitted that the study was a fake, and that the amount of data he had gathered was too small to be meaningful.

Vicary shied away from any public exposure after the uncovering of his scam, but the urban legend of subliminal advertising never died down. Recently, however, scientific breakthroughs in the way our brain works, and the way we react to stimuli, have fueled a completely new wave of interest in the power of the unconscious. And this time, it's not based on some phony experiment in a movie theater.

UNCONSCIOUS BRANDING

According to the founder of Unconscious Branding, Douglas Van Praet, our conscious mind is extremely overrated, and it's really our unconscious mind that does all the heavy lifting. Marketers are quickly latching on to this.[9]

Many processes in the brain occur automatically, fielding an amazing amount of information without any involvement of our conscious mind. This is a natural mechanism to prevent us from becoming overloaded by simple, routine tasks. But when it comes to decisions, we still tend to assume that they are made by our conscious mind. This turns out not to be true.

Timothy Wilson is a professor of psychology at the University of Virginia and the author of *Strangers to Ourselves*. He states that our senses can accumulate more than 11 million bits of information every second, while our conscious mind can process only about 40 bits per second.[10]

Classical studies in experimental psychology have confirmed that the limits of our conscious working memory are at about seven items (plus or minus two). That's why designers tend to put fewer than seven items on a website, why we struggle to remember more than seven items on a grocery list, or why we have difficulty in remembering numbers with more than seven digits. When the brain is overwhelmed with thought, it is forced to juggle more balls than it is capable of juggling.

It turns out that the unconscious part of our brain really runs the body. It controls pretty much all the sensory perception, and all the continuous bodily functions that happen mostly unnoticed. Your unconscious mind, at this very moment, is keeping your balance, commanding your heart to beat, telling your lungs to breathe, growing your hair and nails, replacing the cells in your body, and removing toxins from your bloodstream.

It is constantly monitoring not only your internal state, but also what is happening around you, always alert for potential threats and opportunities, such as predators, food sources, and mating partners. The main goal of our unconscious minds is self-preservation, the survival and replication of our genes and ourselves.

If we want to understand our behavior as consumers, we have to factor in the unconscious aspects of our judgments at least as much as the conscious ones.

FEAR, TRUST, AND HORMONES

Joseph LeDoux is the lead singer of the Amygdaloids, a New York City rock band, whose 2007 debut album was called *Heavy Mental*. Chances are you've never heard of the band, which is entirely made up of researchers in the field of cognitive neuroscience. The band is named after the amygdalae, almond-shaped parts of your brain that have been shown to perform a primary role in the processing of memory and emotional reactions. LeDoux is one of the key neuroscientists doing research in this field.

LeDoux, a professor of neuroscience and psychology at New York University, also directs the Center for the Neuroscience of Fear and Anxiety, which is devoted to using animal research to understand pathological fear and anxiety in humans. He pioneered the study of emotions as a biological phenomenon. His research on anxiety demonstrated that our bodies take action without our minds even knowing. "The conscious brain gets all the attention, but consciousness is but a small part of what the brain does, and it's a slave to everything that works beneath it."[11]

LeDoux's findings basically showed that the amygdalae can literally hijack our mind and body, causing us to respond while completely bypassing the cerebral cortex, where our conscious awareness resides.

Antonio Damasio, professor of neuroscience at the University of Southern California said it beautifully: "Our emotions influence our thinking much more than our thinking influences our emotions." Damasio, who heads up the Brain and Creativity Institute, argues that René Descartes' mistake was the dualist separation of rationality and emotion. (Descartes famously coined the phrase *cogito ergo sum*—"I think, therefore I am." He was a firm believer in the power of the rational brain.) Damasio and LeDoux, four hundred years later, put the emotions and unconscious before the logic.[12]

We now seem to understand that our behavior, our response to stimuli, information, and marketing, isn't as logical or as cerebral as we would expect. We aren't simple beings who can be controlled and manipulated with simple algorithms and rules. We are complex organisms, with a bias toward unconscious processing of inputs. We make decisions and choices based on our intricate systems of chemical reactions. To succeed today, marketers need to understand that.

MARKETS ARE BECOMING NETWORKS

Douglas Van Praet once wrote that "All marketers are in the business of buying and selling good feelings. We are all peddlers of dopamine." Good marketers are selling the best drugs in the world, produced by the exquisite pharmacies of our brains: the molecules of our own feelings.

Dopamine is indeed one of the key ingredients in this process. The midbrain region regulates our levels of motivation and our ability to predict rewards by releasing dopamine in the frontal and temporal regions of the brain. But there's much more than dopamine.

Probably the most important hormone in this regard is oxytocin, also known as the love drug. Oxytocin is most easily released by the brain as a result of physical contact, spiking during activities such as breast feeding, hugging, cuddling, and sexual intercourse.

Dr. Paul Zak of Claremont Graduate University is one of the founders of the emerging field of neuroeconomics. He says that oxytocin is the social glue that binds people and societies, as well as the economic lubricant that enables a host of transactions on which markets depend. As Dr. Zak explains, "We discovered that … oxytocin is released when someone trusts us and induces us to reciprocate trust."

It turns out that oxytocin functions in virtual environments, not just physical encounters. Zak observed that exposing a subject to content, messages, or images using social media can result in a double-digit increase in oxytocin levels. The brain apparently does not seem to differentiate among real, imagined, and virtual friends.

So, what does that mean for marketers? Do they have to study the inner workings of the brain, become experts on the amygdalae, and understand the chemical reactions of oxytocin and dopamine? Buy a copy of *Heavy Mental*? No. But we have to be aware of two hugely important trends that are coming together: markets are becoming networks,

and individuals are responding to inputs based on the concept of neuroeconomics.

I believe the future of marketing is network neuroeconomics. In this era,
companies must understand how to influence networks in order to influence
individual consumers. They must tailor their efforts to individuals who are
unique, yet all of whom are part of a continuous stream of information, and
whose behavior is only partly governed by conscious decisions. They must
understand that marketing has been transformed from a simple linear system to a complex one—a Barabási scale-free network of information flux—
with the consumer at its heart.

THE TRIBE HAS SPOKEN

Seth Godin, the Internet-era marketing guru and bestselling author, talks
about the end of the *TV-industrial complex*. Godin was a software marketer
at first, but then started one of the first Internet-based direct marketing
companies, Yoyodyne, which was acquired by Yahoo! in 1998. He remained
the head of marketing for Yahoo! until 2000.

In the first wave of the Internet, Godin promoted the concept of permis-
sion marketing, where enterprises provide something that is "anticipated,
personal, and relevant," in contrast to advertisements on television and
radio, which were seen by Godin as "interruption marketing" because they
interrupt what the customer was doing.

Recently, Godin has argued that the Internet has ended mass marketing
and revived a human social unit from the distant past: tribes. We seem to be
reliving an ancient organizational model, constantly seeking to assert our
self-interest while still remaining firmly within the good graces of the tribe.[13]

But all of a sudden, the tribes have gotten much bigger. They have scaled
to network size. Today's network behavior, the growth of social networks, for
example, reminds me of Yogi Berra's famous paradox: "Nobody goes there.
It's too crowded."

The world of network neuroeconomics will be ruled by those who understand how to influence a flux consumer in a superfluid network of information. We need to develop a complexity theory for the world of marketing,
where we abandon the idea of absolute control over a target demographic,
but adopt the VUCA model of volatility, uncertainty, complexity, and agility.
We need to understand that markets have become networks, and that we
need to influence dynamic behavior and abandon the simplistic view that
we can see markets as merely first-order systems.

We have to understand how to influence tribes.

THE AGENCY WILL DO ALL THAT

The traditional approach to marketing in the last 50 years has often been to hire an agency and let the people there figure it out. It's complex enough, it's scary enough; let's outsource it to someone who at least has the industry awards to prove that they know what they're doing.

The agency model keeps changing all the time, especially in terms of compensation. Recently, the model seems to have broken down completely, only to be replaced by the equivalent of the agency in the network neuroeconomics era.[14]

When a brand wanted to reach a target group of consumers, it would engage an agency. The agency would craft a clever message, build a brand universe, and figure out which media it would use to reach the target demographics. The choice of media was very much related to the type of technology that would be the right fit: radio, television, billboards, banner ads, mobile ads—whatever would have the best chance of reaching the target group. Typically, these channels are very expensive.

And then the dynamics of the engagement would unfold.

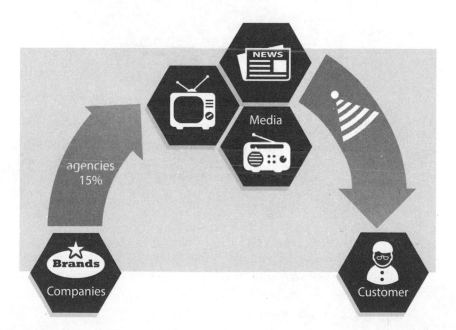

The standard model of media billing was 15 percent, which meant that when the media buying and the creative work were managed by the same agency (as in the classic *Mad Men* era), the agency would receive a 15 percent commission on the media spending, and would then carpet bomb consumers in the media it had chosen, hoping to hit enough innocent bystanders to justify its marketing campaigns. (It's no coincidence that a strategic military operation and a marketing venture are both called *campaigns*.)

But the digital world started to change all that. Things could be measured much more accurately. You could measure how many people followed a link, how long they stayed, how many pages they viewed, and how long they looked at your ad before they clicked away. And that changed things—profoundly.

In the digital-media world, the 15 percent model evaporated, and more and more corporations began to want 100 percent pay for performance, in which the client pays media agencies strictly on a per-action basis (an action being a lead, sale, registration, download, or click).

Today, the model of the 15 percent and carpet bombing has been replaced.

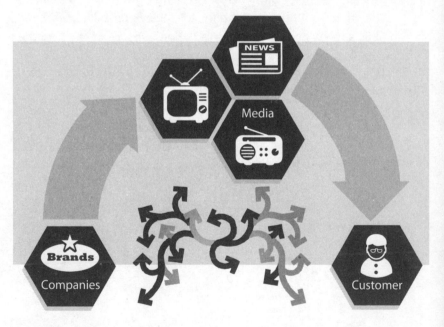

Markets have become networks of information. Brands and channels are only part of the information that flows through the consumer, who is constantly being fed information, triggering dopamine and oxytocin in a way that *Mad Men*–era agencies never imagined.

The problem isn't finding the right channels anymore. The problem is getting through. How do you influence an individual who is so engulfed in information? The old adage *content is king* is long gone. Coherence is the new mantra. The individual has no need for extra information, but is looking for meaning—for patterns.

Our minds are designed to seek patterns. Finding meaningful patterns in a sea of information is the paradigm of the flux consumer.

Marketers, and society at large, have failed miserably at making our lives easier. On the contrary, instead of making them easier and simpler, most marketers have made our lives more difficult.

As Douglas Van Praet says, "The average drugstore sells 350 different kinds of toothpaste and 55 floss alternatives. We are faced with an increasingly cluttered and fragmented media environment of conflicting messages." In the United States, studies have shown that consumers are being exposed to more than 3,000 ads per day. Brands and marketing messages overload consumers and undermine the very purpose of brands as shortcuts to easier choices and better lives, instead creating a lose-lose situation for marketers and consumers alike.

THE PARADOX OF CHOICE

Psychologist Barry Schwarz states that the abundance of choice actually demotivates us to make purchase decisions. When we're overwhelmed, we intuitively search for beacons of trust. And traditional approaches to marketing and branding don't give us that sense of trust any longer.[15]

Instead, we've put our trust in the network. We trust individuals in the network—friends and strangers alike—more than we trust spokespersons and actors on TV. We trust what we hear from the members of our Godin tribes more than we trust the phony messages crafted by Madison Avenue agencies.

Recently we've seen the rise of measuring your personal influence on social networks with tools such as Klout. Klout is still in its infancy, but the trend is clear: where we used to trust a brand or a particular media source, like a newspaper, we're now beginning to trust individuals in the network.

In the past, the effectiveness of those media carpet bombings was analyzed by companies such as Nielsen. In the future, we will have to adopt a different approach based on the influencing of networks, and understand how to approach the nodes of the network that really matter.

CONCLUSION

How should we tackle the concept of marketing in the future, and how should we consummate this marriage between technology and marketing?

What will be long-lived strategies in an increasingly fluid world of customer behavior?

Here are some tips.

1.   Averages Are Out: Beyond Peppers and Rogers

Averages are so last century. Who is your average consumer, anyway? If you were to tell one of your average consumers that you considered her average, she would probably punch you in the face—or at least be horribly offended, irritated, or frustrated.

The direction that Peppers and Rogers started talking about with the market of one was promising. But the way they approached it—as a mass customization rather than true personalization—was wrong. Consumers want to be treated as unique. We have to understand behavior better than ever. We have to take into account both the conscious and the unconscious influencing of that unique individual, and we have to be able to anticipate what you really, really want. You. Not the average demographics that represent you. You.

2.   Listen and Understand

Paul Arden said, "If you want to be interesting, be interested." That's exactly what we need to do in the network neuroeconomics era. The world has become complex, and the network only accelerates that. We can't rely on simple action-response types of behavior. Instead, we have to understand complex, dynamic behavior, and therefore we have to find patterns, seek insights, and embrace complexity.

The irony, in other words, is that, although we want to be treated as unique, we are not because the patterns that emerge in the behaviors of others help organizations predict what we might like. Amazon is able to recommend products tailored to the unique human beings we think we are … but only on the basis of the patterns in the behavior of so many others who act very similarly.

3.   Trust Is a Must

The core fuel of the network neuroeconomy is trust. But building trust has to be genuine, rooted in an understanding of the rules of play in the networked society. You can't control your customer anymore; you can't force a demographic in a certain direction. You have to influence people in a way that won't compromise the trust of the network. As Godin pointed out, "Human beings can't help it: we need to belong. One of the most powerful of our survival mechanisms is to be part of a tribe, to contribute to a group of like-minded people." Those tribes are based on trust.

I like the brave example of Clorox. Its stain app always recommends the best option to use for a stain, even if it is the competition's product. And Hilton advises travelers—even those staying in competing hotels—about the best places in town to eat or drink. That is how you build trust—by being bold.

Today's simplistic advertising models on the Internet are plain horrible. Who hasn't experienced a banner ad on a website that you visit, but you're not interested, so you avoid it—only to be confronted with the exact same ad on the next website you visit, perhaps with different wording or a different picture. You click away again, and, to your horror, the ad just keeps following you like a drunk barfly. No wonder Internet users are becoming completely immune to this simplistic type of advertising—just as bacteria become resistant to antibiotics if you overuse them. Engagement is an earned currency.

4.   Change and Become Multifaceted

"And yet I told Your Holiness that I was no painter"—Michelangelo to Pope Julius II, who was complaining about the progress of the Sistine Chapel ceiling in 1508.

I love this quote from Michelangelo, even if he probably didn't say exactly this to the pope. It took him four long years to paint the ceiling, and he only finished it in 1512. It then took him six years, between 1535 and 1541, to paint *The Last Judgment*, a magnificent fresco on the altar wall of the Sistine Chapel.

Next-generation marketers aren't just about building messages to consumers or understanding consumer behavior. Next-generation marketers have to understand technology—not to the extent that they should be able to code the next wave of social marketing gizmos, but to the extent that they understand the dynamics and mechanisms that rule the world of technology. Next-generation marketers have to understand neuroscience—not that they should be able to carve out the amygdalae, but that they should understand the influencing of individuals based on the emerging neuroeconomics mechanisms. Above all, next-generation marketers should become network thinkers, understanding the influence of network nodes in a world where consumers have become part of an ecosystem of information.

And perhaps that is the most important lesson. The future isn't binary anymore. The world has become warped, intertwined, and mingled. And to make sense of this multidisciplinary world, a multifaceted approach will be vital if an organization is to survive.

THEY BOUGHT A ZOO

YouTube is one of the most important technology companies of the last decade, and one of the core drivers of our networked society. The numbers of views, viewers, clips, and movies on YouTube is staggering. But YouTube is a technology company, run by engineers, fueled by engineers, and driven by engineers.

YouTube did a brilliant thing a while ago. It hired people from the advertising world—dyed-in-the-wool *Mad Men*—and created a concept called the Zoo. The Zoo is run by Mike Yapp, a tremendous creative talent who was creative director at Carat, a huge agency with thousands of employees owned by the Aegis Group, which is listed on the London Stock Exchange. After a very successful career in the advertising world, Mike retired and spent his days pursuing his true passion in life: surfing on California beaches. But then YouTube called and invited him to come and set up an agency inside YouTube that would serve as the bridge between Google, the world of search; YouTube, the world of video; and the world of the advertisers, which was where YouTube made its money.

The Zoo was an experiment, but it was an internal network, free from the constructs of the traditional YouTube hierarchy. It was there to influence the big, engineering-driven company to become more in tune with the world of advertising. The results were spectacular, as the Zoo became the key driver for YouTube, enabling it to build strategic relationships with advertisers as well as a solid business revenue model.

The name *the Zoo* was aptly chosen: it was a wild gang of all sorts of people: engineers and creatives, advertising executives and hackers, sociologists and psychologists, neuroscientists and librarians. Well, maybe not librarians.

The point is, if you want to understand the future of marketing, you need to build your marketing department as a Zoo as well—a Zoo without limits, without boundaries. Marketing isn't a skill anymore, and it's not done well by a department filled with marketers anymore.

In the future, marketing is a network.

WATCHING WATER BOIL

Water boils at 100 degrees Celsius.

But what I find fascinating is that you can heat water up to 99 degrees and you won't see a thing. Water is just water, only hotter. But then, when you reach that magical 100 degrees, all of a sudden the water is bursting with bubbles and vortexes, bristling with excitement and impending danger.

That is happening with markets all around us. We've been heating up the markets, pumping more and more energy into them, with little or no tangible results or visible marks. But then the market suddenly flips, turning into a bubbling network of intelligence.

When markets flip, the consumer suddenly holds the cards. The consumer has become empowered, and has full control. This is the fundamental characteristic of a market flip. The term *consumerization* was used primarily during the advent of the New Normal, when the power of technology, once the domain of the technology elite, came into the hands of consumers.

Instead of being told by the high priests of IT departments which technology they can use and which devices they can have, today users themselves choose which devices they carry. They have the power.

The same consumerization will happen to marketers. When markets flip, the old control exercised by marketers will evaporate. The ability to dictate the conversation with the consumer is fading quickly. Markets have been transformed into networks of intelligence, which follow different rules and observe different behavioral patterns.

To keep up, companies will have to flip as well. They will have to rethink their approach to partnering, to sourcing, and to innovation. Companies will have to flip their internal structures in order to survive and be able to react rapidly enough.

When the markets start to boil—when they flip—they become magically alive.

Be ready.

THE ERA OF
NETWORKED HEALTH

The field of medicine is advancing at breakneck speed, and that pace is only accelerating. The sequencing of the human genome was first completed in April 2003 by Craig Venter and his team in San Diego, a breathtaking leap in human achievement, but at great cost: the total investment made to unravel the human genome was approximately $2.7 billion.

In 1953, Watson and Crick introduced DNA's elegant double helix to the world in the pages of *Nature*. With extravagant understatement, they'd begun their letter to *Nature* by noting: "This structure has novel features, which are of considerable biological interest."[1]

In 2003, anyone with access to the Internet could look at an atlas that contained an entire human genome for the very first time. Only half a century had passed between the discovery of the structure of DNA and the first "reading" of the entire genetic code of a human being.

In 2007, the genome of James Watson, the codiscoverer of DNA's double-helix shape, was sequenced for just around $1 million. In 2013, the typical cost to map out an individual's genome is between $1,000 and $4,000. The cost per bit of biological information is decreasing at a rate faster than Moore's Law, and the race is on to be able to sequence an entire human genome at a $100 price tag.

At the same time that technology is allowing us to understand more about our bodies than ever before at ridiculously low costs, we also see an enormous eagerness to measure our activities and take an obsessive interest in our own bodies. We see people wearing Fitbits and Nike FuelBands, which measure, store, and visualize our core vital signs. But that's merely the beginning.

This movement is often called the "Quantified Self," where people are trying to understand how they function by "self-knowledge through numbers." The idea is to collect all sorts of inputs (cardiac rhythms, food intake, exercise regimes, blood oxygen levels) and correlate this with performance. In fact, with today's technology, this is not only possible, but also quite simple and cheap to perform. We have ample technology that allows us to track our vital signs, store them, send them to the cloud to be analyzed, and interpret and visualize the results.

This could really revolutionize the way we monitor and track our health. In traditional medicine, you would have a doctor investigating your heart once a year. A typical examination would be a minute per year of cardiac observation with your physician holding a stethoscope to your chest. That would be roughly 80 heartbeats per year.

Over an entire lifetime, this would add up to 6,000 heartbeats "heard" by the medical profession. If you consider that a human heart will beat 3 billion times during a full lifetime, that's peanuts! It's statistically irrelevant. But in the age of the Quantified Self, you could have your entire history of heart patterns, blood status, temperature curves, muscle stress, and oxygen flow uploaded and observed every living moment of your life. Not by humans, though. By machines and algorithms.

AVERAGES ARE OUT, EVEN IN HEALTHCARE

Healthcare is going through a series of seismic changes. At the core of this disruption is healthcare's evolution from a value chain to an ecosystem with the patient at its core.

Today, the patient is in charge, and around the patient we find a system of players, including websites, social media inputs, online communities, Google, and Wikipedia. The pivot of the ecosystem is the patient; he is in charge, and he now influences the physician as much as the physician influences him.

Traditional healthcare is a game of averages. When you become ill, the current healthcare model tries to figure out as quickly as possible what is wrong with you and which disease you have, and to provide you with a treatment that works for *most* people who suffer from that condition, based on average responses during clinical trials. There is no guarantee that it will be right for you. It's just an indication that, on average, it's the right treatment.

The thing is, when you are ill, you don't want to be seen as "average." When I'm ill, I want to be treated for my exact condition, using my genetic background and my medical history. I want drugs that are good for *me*. I don't care about averages.

Medicine in the future will be ultrapersonalized. People will want designer drugs. And I don't mean the recreational type. I'm referring to the molecules and compounds that will have the best possible chance of curing their condition.

The scholar Richard A. Epstein argues that today's focus on the "average" response to medicines, which leads to a go/no-go decision, breeds regulatory thinking that overvalues risk, ignores individual differences, and needlessly deprives patients of valuable treatments.[2]

Today, personalization is still an extremely expensive and rare approach. When Steve Jobs was diagnosed with pancreatic cancer, he had a team of researchers in several institutions around the world sequencing his DNA in order to develop a treatment that would target his specifically mutated cell pathways. In 2009, he went for an experimental treatment in Switzerland that was tailored to his exact situation. When he passed away, he was one of the few people in the world to have had his genome sequenced.

But with healthcare advancing so rapidly, such sequencing will no longer be a privilege of the rich; it will eventually become mainstream. Today more than half a million Americans have used services such as 23andMe that offer "rapid genetic testing." The company is named for the 23 pairs of chromosomes in a normal human cell, and it was founded by Anne Wojcicki, who was married to Google cofounder Sergey Brin.

23andMe began offering DNA testing services back in 2007. Customers deliver samples of their saliva to have their DNA analyzed. The results are posted online and allow users to assess their genealogy, inherited traits, and possible congenital risk factors. When you use this service, you can see your likelihood of getting, for example, diabetes, Alzheimer's, or dementia.[13] Healthcare in the future will not be about averages anymore. It will be about unique individuals. But the availability of this technology, combined with the insight into your personal health situation, will raise myriads of moral and ethical dilemmas beyond anything we've ever encountered.

In the future, your genetic code will be imprinted on an ID card. Medicines will be tailored to your genes, and will help prevent specific diseases for which you may be at risk.[4]

Insurance companies' operations are based on the law of averages, on understanding the science of spreading risk over large groups of people. But the business of health insurance will change profoundly as a result of the shift from averages to individuals. Government regulatory reflexes will probably ensure that insurance companies or employers will never be able to force individuals to release such private information.

But what if people *voluntarily* give insurance companies or employers insight into their individual private health records? What if people voluntarily give an insurance company access to their DNA profile, and tell it: "Give me your best deal on life insurance." Health then becomes ultrapersonal.

RUST IS OUT

In the same way that the dental profession has seen an incredible shift in its offering, we will probably see the same in the field of general health-care. Fifty years ago, a dentist was predominantly involved in treating dental problems: pulling teeth that had become infected with decay. Today, with the enormous advances in dental hygiene, the dental profession is involved in helping take care of your teeth, perhaps correcting the alignment of your teeth, but essentially playing a much more important role in preventing dental problems than in curing them.

It reminds me of the many shops that used to provide "antirust" treatments for cars. I grew up in a rather wet part of Europe where cars would suffer severely during cold winters. Cars could be treated against rust, protecting them from the elements, but today those shops are virtually gone. Cars don't rust anymore. We've figured out how to make cars that aren't affected by the seasons.

The advent of monitoring technologies and the abundance of sensors, combined with the ubiquity of communication that can translate health information from your body into algorithms, will allow for a totally new breed of players in the healthcare industry. These companies will provide "health as a service" and will keep our bodies in good shape, providing corrective and proactive care, based on the constant monitoring of our personal health. Instead of pulling teeth and treating rust, these service providers will become true guardians of our health. Erik Topol calls this the age of the "end of illness."[5]

SO, WHOM WILL YOU TRUST?

The "powers that be" in the healthcare industry disagree, and Topol's vision is greatly contested. But the shift to becoming information-rich and permitting the industry to focus on individual patients instead of laws of averages allows for many disruptive new business models that will change the landscape of healthcare. Big time.

This shift will also change our landscape of trust. In the past, patients had blind faith in physicians, hospitals, and drug companies—because they had to. There was no alternative. In today's hyperconnected society, however, where information flows and transparency is the new normal, that trust is being reshaped.

Today, when you need to have major surgery done, it actually makes sense to check out whether the hospital you are going to is really the best one for replacing a knee or a hip, or even for cardiac bypass surgery. Today, the results and success ratios of medical procedures are available online, enabling patients to make educated choices. In a world in which we compare holiday resorts on sites such as tripadvisor.com, we shouldn't expect anything less when we're having our heart valves replaced.

The doctor has always been a person of enormous prestige. Throughout the ages, a man of medicine was a pillar of society, with monumental authority. If the physician recommended that you put leeches on your chest to lower the fever, by God, leeches it was. And if your GP recommended that you take a particular drug to cure your aches, how could you possibly object? The age of networks changes all that in a major way.

The transparency of medical information, success rates of surgeries, and ratings of doctors is just the beginning. More and more people are beginning to share information online, allowing patients to compare. The advent of platforms such as PatientsLikeMe.com infuriated many of the physician lobby groups around the world. This particular site allows people who suffer from an illness to connect to the platform, find people with similar disease patterns, and start engaging, comparing, and exchanging information. What drug did you get? What dosage did you take, and what were the results?

When you have a life-threatening disease, concerns about privacy may become secondary. Above all, you want to find out everything you can that could keep you alive, and you may be willing to share any personal information that could make you or others better. That explains the success of social networks such as PatientsLikeMe.com, and it explains the shift in trust that is taking place. Patients are starting to trust one another, and the network is starting to flow. In an age where increasingly more patients will be able to upload more and more of their personal data, the aggregation of healthcare information will become valuable in the network, and ready to be mined.

Craig Barrett, the former CEO of Intel, sees an analogy between the evolution in the computer industry and the evolution in healthcare. In the early days of IT, we had the mainframe era: long before the advent of personal computers, the world's computing power resided in large machines controlled by technicians. They constituted the elite of specialized mainframe professionals. In today's "cloud world," where the rise of online services and online applications has allowed individuals to decide for themselves which functionalities they want and when, we have migrated toward a self-service model of technology.

In Dr. Barrett's analogy, doctors are the mainframes in today's healthcare world. The coming ehealth revolution will change their role dramatically, with the rise of more personalized, tailored, and self-service healthcare displacing the doctor's role as the key dispenser of information.

When you look at the evolution of the healthcare landscape, and observe these two fundamentals—the spectrum of patients' trust moving from traditional institutions toward networks and communities of patients, and the attitude toward healthcare moving from the traditional reactive health model toward a more proactive one—you have four distinctly different scenarios that can play out.

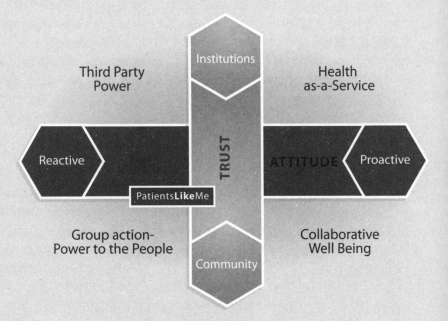

In the traditional corner, the upper left, the institutions (hospitals, big pharma) remain the major source of trust. The focus on a reactive (disease) attitude toward healthcare means that the current model, third-party power, remains dominant. If the attitude were to shift toward a more proactive attitude to health, this could lead to the rise of healthcare providers who offer health as a service to their communities of customers.

When we look at the bottom scenarios, where the power of trust shifts toward the network, and where the communities of trust prevail, we see that current platforms such as PatientsLikeMe.com are just the beginning of a model of collective health group action. Within that, a more proactive approach toward health could result in business models involving collective and collaborative well-being.

The more likely scenario is that the world of healthcare will be confronted with all of these different models acting out their roles, making the healthcare landscape richer and more challenging than it has ever been. But the advent of technology and the rise of networks have the power to completely reshuffle the world of health toward an era of networked health.

THE TRICORDER

The X Prize Foundation is an inspiring organization that routinely launches impossible goals that are rewarded with huge prizes for whoever reaches them. In 2004, a group of wealthy Silicon Valley billionaires launched the foundation by promising a reward of $10 million for the company that could make commercial space flight possible—that is, the first nongovernmental organization to launch a reusable manned spacecraft into space twice within two weeks.

The philosophy behind the X Prize is related to the story of Charles Lindbergh when he crossed the Atlantic Ocean with his plane, *The Spirit of St. Louis*, in 1927. Lindbergh performed the first solo, nonstop flight from Roosevelt Field in Garden City, on New York's Long Island, to Le Bourget Field in Paris, France. He nearly didn't make it, and it was a major achievement at that time. But the real reason that Lindbergh flew across the Atlantic in brutal conditions was that he wanted to win a reward. The prize was $25,000, a fortune in those days, awarded by the French-born hotel owner Raymond Orteig,[6] who was fascinated by air travel. Lindbergh was crazy enough to risk life and limb, and this gave birth to the intercontinental commercial air travel that we take for granted today.

The X Prize Space Challenge was launched for exactly the same reason. For a $10 million prize, dozens of companies (including Virgin Galactic, owned by the eccentric billionaire entrepreneur Richard Branson) sprung up to compete for it. The result was the creation of a multibillion-dollar commercial space industry. Another contestant was Space-X (the brainchild of the equally unconventional and charismatic billionaire entrepreneur Elon Musk). Space-X received a $3 billion contract with NASA to replace the space shuttle with its solution. The prize was won on October 4, 2004—the forty-seventh anniversary of the *Sputnik 1* launch—by the Tier One project designed by Burt Rutan and financed by Microsoft cofounder Paul Allen.

As the X Prize Foundation states: "If the government had given $10 million in government subsidies, we would have received in the end three whitepapers and a PowerPoint on the Future of Space Travel. Now we created a multibillion-dollar industry with a $10 million prize."

In 2011, the X Prize Foundation launched a new challenge, the Tricorder. The name comes from the device from *Star Trek* that could be used to diagnose ailments instantly. The ship's medical officer would whip out his medical tricorder and instantly figure out what was wrong with the patient. The $10 million Tricorder X Prize competition, sponsored by Qualcomm, has dozens of companies around the world that are trying to build a device that combines ultraconnectivity with ultrapersonalization to deliver a medical instrument built for the age of networks. As one of the contestants, a start-up called Scanadu, states: "We'll put your Smartphone to MedSchool."[7]

We are truly the last generation that will know so little about our health. In the age of networks, the flow of health information is bound to be awe-inspiring. Just as the lifestyle of our grandparents seems primitive to us, today's medicine will look like voodoo to your grandchildren.

CHAPTER 6

WHEN ORGANIZATIONS BECOME NETWORKS OF INNOVATION

This is the first generation to grow up completely connected to everyone and everything. That is what makes them truly unique. Millennials should really be called the Network Generation: Generation N.

Static corporate structures and linear career paths will have to be reinvented for the age of networks. To thrive in a network-based society, companies must become more fluid.

My father worked for the same corporation, the largest oil company in the world, for his entire career. He was recruited by this corporation the day he graduated, and he worked there until he retired. My father loved his job and respected his employer. He didn't always like the structure he worked for, though. Many nights I remember him coming home complaining about corporate politics, silly rules, and incompetent managers. That's probably the reason I became an entrepreneur.

But the relationship between my father and his employer was characterized by one word: *loyalty*. The employee was loyal to his employer, and in return, the employer took care of the employee. That's when everything was rosy.

The concept of a *career* meant something in those days. It was a path, and the vector of the path was upward. You would start at the bottom and work your way up. And every year, just as you measure the growth of your children against the doorpost, HR would put you up against the wall and measure how much progress you'd made up the corporate ladder.

A career was signed off on with a contract that sealed the trust between you and the employer. The company wanted to be able to trust that you would do your best. In return, the company would look after you—supply you with a desk, a computer, a phone, pencils, a car, a parking space, and buckets of coffee to get you through those dreary meetings. Sounds like a fair deal.

And, of course, you would be entitled to *compensation*, which the dictionary describes as "The act of giving someone something, such as money, as payment for a ... loss." So compensation was money you got for the loss of what, exactly? The freedom you gave up when you joined the company?

I'm not trying to dis the structures and mechanisms of the past. I think they worked brilliantly when we had companies that were built for executing plans and were designed for efficiency and scale, and when we didn't have the technology of the New Normal—the age before network thinking had become the prevailing paradigm.

GENERATION N

But today's reality is completely different. By 2020, 46 percent of the U.S. workforce will be millennials, or "generation me me me," as *Time* magazine called them. And this generation has network thinking built into its very soul. In the past, I made the mistake of thinking that what really defined this generation was its capacity to deal with all things digital. I was wrong. It is true that these people have great skills with anything that has an on/off switch, but it's much deeper than that. This is the first generation to grow up completely connected to everyone and everything. That is what makes these people truly unique. Millennials should really be called the Network Generation: Generation N.

This next generation has grown up on Tumblr, Facebook, and Twitter.

It understands the rules and dynamics of the network.

The network doesn't care about hierarchy. In a world that has shrunk to six degrees of separation, why should I care about a structure and a hierarchy that has 10 managers between me and the CEO? Those in Generation N have learned to live in the network, make connections in the network, and advance through the network. For them, it's not about *up*. It's not horizontal. For them, it's all in the network. It has become a level playing field. The vector points to the network.

If you want to advance in the network, you have to feed the network. You have to share information and help accelerate it in the network. Imagine the culture clash that will take place when millennials enter a workforce in which information is still power—where it's hoarded like a good case of wine that you keep in the cellar.

The network is a pure meritocracy. According to Gary Hamel, if you post something on YouTube, people don't ask you if you went to film school. If it's a great video, it's a great video. You earn your rights in the network based on what you do, what you provide, and what you share, not on some badge of honor.[1]

And that will truly change the way we work, the way we build organizations for the future, and the way we think about companies.

I'm always amazed at how many companies today are trying to implement a new world of work inside their organizations by completely focusing on the physical and practical side of things. They tear away the cubicles to create open spaces. Great. They give everyone a cool laptop and a smartphone. Great. They implement flexible working environments, set up cozy meeting places, put really cool design furniture into the cafeteria, and bring in a snooker table or a vintage arcade game in the coffee corner. Great.

And then nothing else happens. It's as though they think that dressing up their offices like the Google campus in Mountain View, California, will transform them into the most innovative company ever. You can dress up your CEO in sneakers, blue jeans, and a black turtleneck and tell him to stop shaving, but that won't turn him into Steve Jobs, either.

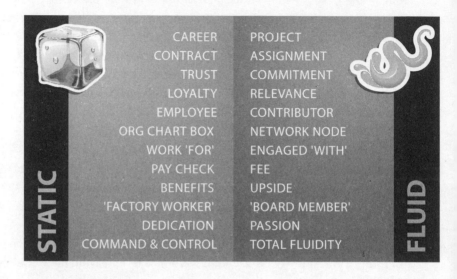

STATIC		FLUID
	CAREER	PROJECT
	CONTRACT	ASSIGNMENT
	TRUST	COMMITMENT
	LOYALTY	RELEVANCE
	EMPLOYEE	CONTRIBUTOR
	ORG CHART BOX	NETWORK NODE
	WORK 'FOR'	ENGAGED 'WITH'
	PAY CHECK	FEE
	BENEFITS	UPSIDE
	'FACTORY WORKER'	'BOARD MEMBER'
	DEDICATION	PASSION
	COMMAND & CONTROL	TOTAL FLUIDITY

ORG CHARTS AND THEIR DISCONTENTS

The new world of work will require a complete rethinking of how people are engaged with your company. We might have to throw away many, if not all, of the tools we used in the past, when companies were still based on that ancient model of the hierarchical structure. Let's start by questioning the most traditional of all models: the org chart.

Org charts are evil. That's the motto of all start-ups, but the org chart is probably seeing its last days in established companies as well. In many companies, it is still the mechanism by which we place people in an organization. It cements someone in a place somewhere in the company, and it shows the lines of reporting. However, it shows only the vertical component of an organization. It totally ignores the network inside a company. It turns meritocracy into aristocracy.

Many people are locked into a job description, labeled from the day they entered an organization. Oh, you're Marketing, IT, or Finance. And these labels, like tattoos, never wear off. Many people have no freedom to alter their label, or escape the curse of being boxed into an org chart.

Yet many human resources departments still cling to these charts. I often feel bad for HR executives. They can seem to be totally unequipped to deal with the challenges of the network generation, since they are using instruments that stem from the industrial age. Many HR tools are remnants of the days of command-and-control structures, and it seems as if HR hasn't been able to reinvent itself, at least not with the instruments at its disposal. In many organizations, HR stands for Horribly Reactive. It will be interesting to see what happens when the millennial tsunami hits the workforce.

Luckily, there are projects and organizations that have ditched their org charts. One of the most beautiful examples is that of Semco. It used to be a company run in a traditional autocratic management style. But that changed completely in 1980, when Antonio Semler resigned as CEO and his son Ricardo Semler took over. On his very first day, he is said to have fired 60 percent of Semco's top management.[2] The company culture and organizational design are based on three core principles: employee participation in management decisions, companywide profit sharing, and an open information culture.[3] And it works: under Ricardo's leadership, Semco's revenue has grown from $4 million in 1982 to $212 million in 2003. No wonder Ricardo's innovative business management policies are widely documented around the world.[4]

But this flat approach doesn't just work wonders in organizations; scientific projects have thrived in the absence of autocratic structures as well. One of the largest and most challenging projects in science—ATLAS, the particle physics experiment being performed with a special detector of the same name in the Large Hadron Collider located at CERN—was the result of a collaboration by about 4,000 physicists, engineers, and computer scientists from 175 universities and research centers in 38 different countries. Suffice it to say that this kind of large-scale and dispersed collaboration needed an innovative organizational structure. It was based on a nonhierarchical system (without rigid org charts) led by spokespersons and technical coordinators. This flat approach indeed proved very efficient in bringing the different cultures together and ensuring a smooth collaboration and communication. And—as we all know—one of science's greatest breakthrough innovations was a result.[5]

REIMAGINING CAREERS

I don't believe that the concept of careers will prevail. Careers are mechanisms of advancement, and the concept of climbing ladders and structures has become rather meaningless in the age of networks. More and more people are focused on individual projects, not on one overarching path. People want to be able to sample new insights and try out fresh approaches. They want to keep their options open and be able to change direction. Old-style serial résumés, which depict one job after another, are on the way out. Today, more and more people are embarking on parallel paths, with multiple interests and simultaneous projects. Valve, for instance, does not offer promotions in the strict sense of the word, only new projects.[6]

I don't believe employment contracts will remain the same. In the past, we asked people, "Whom do you work for?" Today, we ask, "What are you working on?" The contract that sealed the trust between employee and employer is being replaced by much more fluid engagements. Today, people are engaged with, involved with, or "connected to" an organization. They won't *work for* anymore.

Independent board members already operate this way. They have a stake in the company, and they may even help to steer it, but the relationship isn't exclusive. People can be on the boards of multiple companies. As a matter of fact, the best board members are often those who are the most connected—those who serve on multiple other boards.

The boards of Fortune 500 companies are populated with extremely well-connected people—people who understood networking long before it was enabled by technology; people who learned networking in school, often at Ivy League universities. Members of fraternities, service clubs, and country clubs, they have long realized the value of networking in their professional, social, and financial affairs. As they used to say, "It's not what you know, it's whom you know."

RELEVANCE REPLACES LOYALTY

In the age of networks, people won't work for a company. They will engage with organizations—often multiple organizations. They will be just as committed, and just as involved, as the board members of a multinational would be. They won't want to work in a box in an org chart. They will want to make themselves relevant for an organization by being involved in the different communities that make up that organization and sharing their skills and competencies.

The static structures of today will have to be reinvented for the age of fluidity. This is a completely new paradigm. Careers will have to be reinvented and replaced by much more fluid thinking. The way we reward people for their efforts will also have to be completely overturned.

The fundamental shift is that we're replacing the notion of loyalty with that of relevance. The company structures of the past were completely based on loyalty mechanisms.

However, the era of fluidity is based on relevance. If you engage with a company on a project, you will have to be relevant, or you won't last. Likewise, a company will have to remain relevant to the people it engages with. Loyalty was a mechanism of command-and-control structures borrowed from the military. It's probably necessary if you're out fighting a war. But in today's VUCA world of network thinking, we have to take it to a new level: the network level.

Cities form a very interesting analogy. When we look at companies today, there is ample statistical evidence showing that the bigger companies become, the more likely they are to experience a dramatic drop in productivity per employee. As companies grow, bureaucracy and hierarchy sap individual productivity. If a company triples its size, its productivity drops by up to 50 percent.

Cities, on the other hand, are completely different. The bigger a city is, the more innovation, the more entrepreneurship, the more jazz clubs and ballet dancers and violinists and artists and potential Nobel Prize winners it has. Why? Because cities are not exclusively hierarchically controlled. Of course there are city governments and fire departments and police precincts. But cities also have networks, which only get stronger as the city becomes bigger. Cities are built on vibrant bottom-up, network-based organizations.

We have built our companies to be like machines that function in accordance with to a fixed set of rules, like org charts. We have to abandon this approach. To thrive in a network-based society, companies will have to become networks themselves.

Scary though the absence of titles and ranks might sound to traditional leaders, it can be implemented, and successfully. Some medium-sized and even very large companies have almost no formal hierarchy. They have pushed out the traditional functions of a manager (planning, coordinating, controlling, staffing, and directing) to all participants in the organization instead of keeping them in the hands of a select few.

The $700 million tomato processing company Morning Star, one of the best known examples of the genre, has no managers whatsoever. Its employees organize themselves by means of "self-managing professionals who initiate communication and coordination of their activities with fellow colleagues, customers, suppliers and fellow industry participants, absent directives from others."[7] To put it simply: if you need material, you buy it; if you need extra talent, you hire it. But the responsibility is on you as well, of course.

For years, General Electric has run some of its aviation-manufacturing facilities with no foremen or shop-floor bosses. Zappos (which has 1,500 employees) and Medium, Twitter co-founder Evan Williams's company, are managed as a *holacracy*, a "management structure based on the tasks a company needs to accomplish, rather than a standard reporting structure."[8] There is no micromanagement. There are no titles. Pixar also works in a flat environment; it is free from the thick layers of formal management and encourages tight collaborations among teams and departments in order to ensure creativity and innovation. All employees are treated as equally important, and executive are not involved in the day-to-day running of the organization.[9]

I love the way the games software company Valve (with about 400 staff members) describes its flat philosophy in its handbook for new employees, which is subtitled "A fearless adventure in knowing what to do when no one's there telling you what to do." In the glossary at the end, cofounder Gabe Newell is hilariously described like this: "Of all the people at this company who aren't your boss, Gabe is the MOST not your boss, if you get what we're saying."[10]

THE ANTIHIERARCHY

I can imagine that this revolution might sound terrible to some—especially if they're in HR. But I believe that we are now seeing many people thinking along the same lines, and the vector points toward the idea of networks.

One of the companies that I admire intensely is LinkedIn, the social network for professionals. Officially launched in 2003, it reached 200 million users in some 200 countries just 10 years later. Not only is it one of the most successful social networks, but it is also gathering an enormous amount of information on how people behave as a network, and how they plan out their professional lives.

The insight that a company like LinkedIn has into organizational behavior is mind-blowing. At large companies staffed by skilled professionals for whom information is key, more than 85 percent of those employees will be on LinkedIn. And on LinkedIn, people not only will report what they do in an organization, but will connect with peers and colleagues, exchange information, search for talent, look for jobs, and start to jazz up their résumé if they're thinking about jumping ship. The insights that a network like LinkedIn has are much more powerful than any knowledge that an HR department has about its own employees. LinkedIn beats SAP, Oracle, and PeopleSoft any day of the week. An HR department will have a static view of a company's current situation. But the dynamic view of a network like LinkedIn will be able to show where individuals are heading. As Wayne Gretzky said, "I skate to where the puck is going to be, not where it has been."

The founder of LinkedIn is Reid Hoffman, who was also instrumental at helping to start PayPal. It is his firm belief that you can't rely on a company to take care of your career anymore. You have to take matters into your own hands. Reid frequently quotes Muhammad Yunus, the Nobel Peace Prize winner and pioneer of microfinance, who said, "All human beings are entrepreneurs. When we were in the caves, we were all self-employed, finding our food, feeding ourselves. That's how human history began, but as civilization came, we suppressed it and we became 'labor.' We forgot that we are entrepreneurs."[11]

As entrepreneurs, we have to be in charge of our own careers. Organizations ought to be meritocracies, with employees themselves—not their titles or their position on the ladder—proving how relevant they are. At the multibillion-dollar W. L. Gore Company, which has a staff of about 10.000 people, employees take on leadership roles based on their ability to "gain the respect of peers and to attract followers," as CEO Kelly puts it. Though it has kept some form of hierarchy (but not in the traditional sense of the word), the W. L. Gore Company is a very flat environment for such a large organization.

LinkedIn founder Hoffman says that rediscovering our entrepreneurial instincts is necessary if we want to stay relevant. He asserts that we should constantly be ready to adapt to ever-changing work environments, and that we should prioritize learning (soft assets) over compensation (hard assets) for the majority of our professional lives. Of course, the network plays a vital role in this. Hoffman says that the power of an individual is raised exponentially by the power of a network. He should know—he runs the biggest professional network on earth!

THE OPPOSITE OF FRAGILE

Of course the network is important. Of course we constantly have to adapt, and of course we all have to leverage our skills in order to stay relevant. The problem remains: What will organizations look like in the age of the network?

The feedback I often get when talking about this concept is that if we all became nodes of the network and abandoned the concept of command-and-control hierarchies in our companies, we would fall into a state of complete chaos.

Such responses are based on our conditioning to see chaos as the opposite of hierarchy. But we don't actually need to fight chaos, according to the brilliant Lebanese-American author Nassim Nicholas Taleb. His question is a very simple one: What is the opposite of fragile? After the near-collapse of the world economy in 2008, it became painfully obvious that many concepts and constructs that we had implemented in society were very fragile, meaning that they are extremely vulnerable to the shocks and stress that a black swan (what Taleb calls rare and unpredictable events that have an extreme impact) event brings. Banks proved to be very fragile. The euro proved to be very fragile. Stock markets are notoriously fragile.[12]

When we try to make these things less fragile, Taleb observed, we often make them so rigid, so static, and so vast that they no longer work. We could make a bank so completely risk-averse that, although it would be less vulnerable, it would not be able to function adequately. So Taleb set out to look for concepts that are antifragile—"things that benefit from shocks, that thrive and grow when exposed to volatility, randomness, disorder, and stressors and love adventure, risk, and uncertainty."

His basic conclusion is that, in our vocabulary, we don't actually have a word for this. We immediately associate the opposite of *fragile* with *rigid*, but, as he puts it, "There is no word for the exact opposite of fragile. Let us call it antifragile. Antifragility is beyond resilience or robustness. The resilient resists shocks and stays the same; the antifragile gets better."

Taleb explores many things in his book that are antifragile, including many examples from nature. Nature has a tendency to build things that are antifragile. Evolution uses disorder to grow stronger.

I love the very practical Netflix interpretation of antifragility. It deliberately and "unexpectedly" sabotages its own systems with the Chaos Monkey algorithm, which kills processes randomly. It haphazardly disables production instances to make sure that they can survive common types of failure without any customer impact. So, by being constantly under fire, its processes grow stronger. And the more frequently failure occurs, the more the organisation learns. It's a really smart self-inflicted fail fast approach.[13] Google X does something similar with its "Rapid Evaluation Team," which basically does "everything humanly and technologically possible to make new ideas fall apart."[14]

I believe we can directly apply this to the way organizations must adapt to the world of network thinking. We have to reinvent organizations as anti-hierarchy, just as Taleb shows us the concept of antifragility. The network can become the foundation for these antihierarchy organizations.

Do we have to choose between the network and the hierarchy? Between chaos and command and control? Perhaps what we need to do is fully realize that instead of choosing, we have to combine. Perhaps, in a spirit of reconciliation, we can be both at the same time.

BOTH WAVE AND PARTICLE

One of the longest-standing arguments in the world of physics is about the nature of light. Is light a wave propagating through space, or is it a stream of particles that makes up a beam of light?

In the seventeenth century, the battle heated up when Dutch scientist Christiaan Huygens argued that light essentially consisted of waves, whereas Isaac Newton decisively argued that light was essentially composed of particles. In today's world, Newton and Huygens would have clashed in an epic flame war on the Internet.

The battle raged on, and by the end of the nineteenth century, the wave camp seemed to have won. With the experiments of Thomas Young and the theory of James Clerk Maxwell, all odds were in favor of the wave guys. But just when everyone thought that the particle theory was dead and buried, it rose from the grave.

None other than Albert Einstein was to prove, in 1905, with a concept called the photoelectric effect, that light was indeed made of particles. When Einstein received his Nobel Prize in 1921, it would not be for his more difficult and mathematically ingenious theory of relativity, but for the simple, yet totally revolutionary, suggestion of light quanta: light as particles that would later be known as photons.

So who won? Both did. In the end, the physics community could decisively prove that light acts as both a wave and a particle simultaneously. This basically ended the era of Newtonian physics, in which we could explain pretty much everything in the universe with Newton's laws. It opened the era of quantum physics, where the very nature of physics became totally different once we observed it in the realm of the infinitely small, like the world of photons, the particles of light.

Where does that leave organizations?

I believe that as our organizations enter the era of networks and fluidity, we have to abandon the concept that the opposite of hierarchy is chaos. We have to adopt a Taleb mentality and consider what antihierarchy would look like. We probably have to accept that organizations will have to act as structures and as networks simultaneously. The opposite of hierarchy is not chaos. The antihierarchy is a concept that I would call the *fluid organization*.

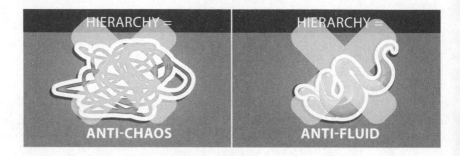

In some circumstances, organizations will have to function as structures, where the command-and-control mechanisms can help them focus on execution and efficiency, driving the organization forward. But in other circumstances, organizations will have to function completely as networks, where the fluidity of the network shines through. They should be both, at the same time. They should be fluid.

In the era of fluidity, we have to adopt this structure-network duality. We have to build organizations that are capable of behaving like a structure and a network at the same time. When we need to drive for efficiency, we need to push on the structure side of the organization, and when we need to drive for innovation and creativity, we need to push on the network side.

If you ask me why Polaroid[15] didn't survive, I believe that while Dr. Land[16] was still running his company, he was capable of maintaining the two sides of the equation: the structure and the network. Sure, he had the organization, the hierarchy, and the structure to run Polaroid as a global business. But at the same time, he nurtured the network of talent, the connections of creative skills and nodes of innovation that kept Polaroid at the forefront of its field. But when Dr. Land was ousted, and new management came in, all it could see was the org chart. Half of the equation, and perhaps the most vital one in a market where you have to focus on innovation, was gone. And a company that becomes blind to its network doesn't survive for very long.

"I usually solve problems by letting them devour me," Kafka once said. Today, many companies seem to implement Kafkaesque organizational structures that are completely ignorant of their inner network. They suffer from org chart myopia. I think the HR departments of tomorrow should shed their Horribly Reactive mentalities and nurture the network within.

Just as markets are disappearing and being replaced by networks of information, with the customer at the heart of the network, organizations need to become networks of innovation. If the outside becomes a network, the inside must become fluid.

Why is that? Because of Conway's Law.

CONWAY'S LAW

Melvin Conway was involved with computers in the very early days, back when they still used punch cards to store computer programs. In his long and fruitful career, he saw a lot of IT projects fail miserably. By observing these failures, he stumbled upon a curious observation.

He saw that when two teams of technology specialists had to work together, the outcome of their work was often a reflection of the social structure of the teams. He phrased it as follows: "Organizations which design systems are constrained to produce designs which are copies of the communication structures of these organizations."

The idea was based on the reasoning that in order for two separate systems to interface correctly, the designers and implementers of each module must communicate with one another. Therefore, the interface structure of a software system will reflect the social structure of the organizations that produced it. And that can prove to be completely disastrous.

A striking example is the sorry fate of the Mars Climate Orbiter, a robotic space probe launched by NASA in 1998 to study the Martian climate and atmosphere. That never happened, because about a year after the launch, when the spacecraft was nearing Mars, it disintegrated into the Martian atmosphere and crashed. The reason was that the teams in NASA and Lockheed that had collaborated on the software had failed to realize that one group had developed the software using metric units (kilograms and meters), and the other group had developed the software using U.S. standards (pounds and feet). Ouch. The cost of the mission was $327.6 million, which by any standards makes for a pretty damn expensive fireworks show in the Martian sky.

Mel Conway submitted his findings to the *Harvard Business Review* in 1967, but the *HBR* rejected it because Conway had not conclusively proved his thesis. Recently, many researchers have found similar evidence of Conway's Law, and even now are stating the reverse: "If you want to achieve success in designing the correct systems, you need to ensure that the organization building it is compatible with the [product] architecture."

In other words, the fabric of an organization (how teams work together) helps determine the fabric of the products it is working on (how products work together).

I believe that we can expand this to the world of networking. As the outside world starts to exhibit more network-like behavior, the inside of an organization should adopt more network-like conduct as well.

Or, to put it simply, if markets become networks, companies will have to become networks as well. And even that might not be enough, because it's not that black-and-white.

BUILD FLUID ORGANIZATIONS

There is a holy trinity at play here, formed by the organization, its structure, and its culture.

There is nothing as elusive as corporate culture. Everyone can feel a corporate culture—almost the minute you walk through the door. But we can't define it, and we can hardly change it. To the chagrin of the millions of executives who have read management self-help books on altering your corporate culture, very few have actually been able to achieve that goal.

I believe that corporate culture is one of the strongest forces in the world of economics, and one of the most powerful engines of growth, but it is almost impossible to get a grip on. But it is definitely tied to the way we structure companies.

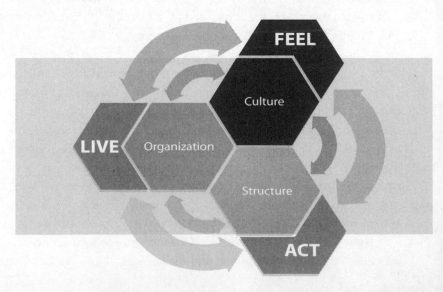

When we make the transition into the era of networks, and reinvent our companies to adopt the dual wave-particle fluidity that will be necessary if they are to survive, we will have to build fluid organizations, with fluid structures and fluid cultures. Because the network always wins.

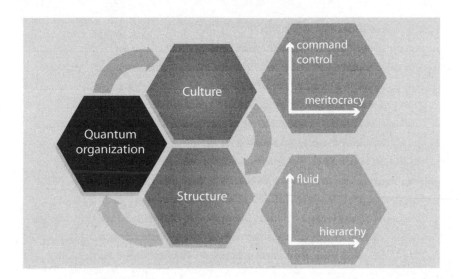

The pharmaceutical industry is one of those for which being able to be both fluid and frozen is essential. It is one of the most regulated and bureaucratic sectors, but simultaneously, it is one of the most innovative sectors and stands to win the most from networked collaboration and open innovation.

Johnson & Johnson, for instance collaborates in an open and transparent manner with a broad ecosystem of scientists and entrepreneurs at universities, academic institutes and start-up biotech companies.[17] Eli Lilly created a spin-off, InnoCentive, a cloud-based innovation management platform. It helps crowdsource innovation solutions and solves problems faster, more cost-effectively, and with less risk than ever before. Since 2001, leading (and sometimes competing) organizations such as AstraZeneca, Booz Allen Hamilton, Cleveland Clinic, NASA, Nature Publishing Group, Procter & Gamble, *Scientific American*, Syngenta, the *Economist*, Thomson Reuters, the Department of Defense, and several government agencies in the United States and Europe have partnered with InnoCentive to speed up their innovation. More than 1,650 external challenges have reached a solving community of more than 300,000, in nearly 200 countries. The success rate of the premium challenges is no less than 85 percent. [18]

CHAPTER 7

CREATION AND DESTRUCTION

You can clearly see the phoenix from
the ashes. Netscape created an entire
industry of Internet companies. Without the
Netscape effect, there would be no Google,
no Twitter, no Facebook.

Innovation requires ongoing destruction. Instead of trying to build indestructible machines that run forever, we should create organizations that foster innovation even after they die.

BROKEN HALOS

Jim Collins and Jerry Porras have sought to understand what makes a company visionary, and what characteristics would guarantee that it was built to last.[1] After six years of research, they published a book about it, which became an instant bestseller in the management book category. Jim Collins became a celebrity on the lecture circuit, and made a fortune. But many of the 18 companies that were portrayed in the book haven't fared so well. As a matter of fact, many of them have been stumbling badly in recent years, and the term *built to last* seems quite awkward in hindsight.

Many people have criticized the book and even the whole genre of books that tries to identify the unique characteristics that make company successful. One of the strongest opponents of *Built to Last* was Phil Rosenzweig, who debunked many of its myths. He described the tendency of experts to point purely to the strong financial performance of a successful company and then to spread the magnificent glow of that performance to the company's other attributes—its clear strategy, strong values, or brilliant leadership. But, as Rosenzweig illustrates, the experts are often plain wrong.[2]

The halo effect is a common phenomenon in psychology. It's a cognitive bias in which one's judgment of someone's character is influenced by that person's other qualities. Physical attractiveness, for example, has been found to produce a strong halo effect: when we find someone physically attractive, we're more likely to perceive that person as being kind, intelligent, trustworthy, or friendly.

The same effect applies to companies: when a company does well on the stock exchange, or has massive earnings and top-line growth, it *must* have the greatest management team in the world, the best leadership in the world, and the best set of values in the world. Often, that is not the case.

BUILT TO LAST

I have had the good fortune of working for many corporations across many industries in my career, both as a technologist and as a consultant. When

I was younger, I had the notion that the bigger a company was, the more efficient and sophisticated it would be.

After almost 25 years in business, and having observed hundreds of companies from the inside, I can safely say that some of the largest corporations are totally clueless about where to go next, suffer from horrible internal politics and inert boardroom cultures, and have given up the hope of transforming their bureaucracies. But when the outside world reads about such companies in business magazines or other such media, they are often revered. And when they stumble or fall, as they all inevitably do, everyone claims to have suspected the dirty dealings, internal bickering, and haphazard strategy all along.

That's why it's really difficult to make a book like *Built to Last*, well, last.

Some of the choices that Jim and Jerry made look downright silly today. Motorola was acquired by Google in 2012. And even companies like Hewlett-Packard, Sony, and Boeing have struggled in the last couple of years, with scandals and rotating CEOs and difficult market conditions.

But perhaps we're asking the wrong question. Maybe it's impossible to remain on top of your game forever. Perhaps there will never be another truly built-to-last company. Perhaps we should spend more time looking at why companies fail, and understanding how that happened, rather than seeking magical qualities that guarantee success.

The more pertinent question today might be: Which companies are doomed to fail? But I doubt whether that book would sell a lot of copies.

CREATIVE DESTRUCTION

If you look at the wonderful ecosystem of Silicon Valley, which is unique in the world, no matter how many regions around the world have tried to copy it, you find that it's one of the healthiest economic ecosystems fueled by death. Silicon Valley is built on two fundamental principles, each of which is all about decay:

1. Schumpeter's ideas regarding creative destruction
2. The concept of the phoenix rising from the ashes, referring to cyclical regeneration

Let's explore.

Joseph Schumpeter was an extremely industrious and ambitious man. If he'd had a LinkedIn profile, you would have had to scroll down *a lot*. Schumpeter was born in Austria-Hungary in 1883, and studied law in Vienna. After getting his PhD, he became a professor of economics and government affairs at the University of Czernowitz in Ukraine. After moving to the University of Graz in Austria, he became the Austrian minister of finance in 1919. Then, in 1920, he became the president of a private bank, the Biedermann Bank in Austria. That bank, along with a great part of the regional economy, collapsed in 1924, leaving Schumpeter completely bankrupt. That wouldn't look good on his LinkedIn profile.

But Schumpeter didn't give up. He became a professor of economics in Bonn, Germany, and then moved to Harvard in 1927, finally moving to the United States in 1932, and eventually becoming a U.S. citizen. Impressive as his CV was, his students didn't really consider him a very good classroom teacher, and many of his academic colleagues didn't like him either, because most of them didn't really understand his ideas. Schumpeter's ideas on business cycles and economic development couldn't really be expressed in the mathematics of his day; the whole mathematical concept of nonlinear dynamic systems was yet to be developed.

But despite his lack of recognition, Schumpeter was extremely ambitious. He famously proclaimed that he had set himself three goals early in life: to be the greatest economist in the world, to be the best horseman in Austria, and to be the greatest lover in all of Vienna. At the end of his life, he said that he'd reached two of his goals (not specifying which ones, although he did say that there were too many excellent horsemen in Austria for him to flourish in all his aspirations).

Today, Schumpeter is most widely associated with the concept of creative destruction, a term that he actually borrowed from Karl Marx, another fine economist with an impressive LinkedIn profile. Schumpeter propagated the idea that in the world of capitalism, innovation in the form of entrepreneurship is the disruptive force that sustains economic growth, even as it destroys the value of established companies.[3]

In other words: it is the task of the start-ups and entrepreneurs of this world to attack the established companies. Although this might topple the traditional players in a market, the end result is that the economy will regenerate and grow.

Many of these economic ideas also have been influenced by Eastern mysticism, in particular the Hindu god Shiva, who is known as being both the destroyer and the creator, combined in one god.

Silicon Valley would have brought tears to the eyes of Joseph Schumpeter, because there, just south of San Francisco, his theory thrives like never before. In Silicon Valley and San Francisco, there are, at any given time, legions of entrepreneurs aiming at killing the previous generation. Inside

Facebook, there are scores of brilliant minds who want to quit and become the next Facebook. Within Google, there are legions of innovators waiting for their chance to cross the road and start a company that will one day leave Google gasping for air. They should all have a picture of Schumpeter tattooed on their upper arms.

PHOENIX FROM THE ASHES

But the second big driver of Silicon Valley might be equally important. From the ashes of failed attempts, new things can flourish. The phoenix, of course, is a mythical bird that is cyclically regenerated or reborn. That idea is at the heart of Silicon Valley. One of my favorite companies in the history of the Valley is also one of its biggest failures: Netscape.

For those of you who don't remember, or are too young to know, that part of history, Netscape was the company that made the World Wide Web really explode onto the scene in 1995. It introduced the browser to the world, became one of the hottest companies on the planet, had one of the most amazing IPOs ever on Nasdaq, and then totally collapsed. But from its ashes, a whole new generation of companies, entrepreneurs, and venture capitalists was able to rise and become more powerful than Netscape had ever been. Obi-Wan Kenobi said the immortal words, "If you strike me down, I shall become more powerful than you can possibly imagine."

The genius behind Netscape was a young software engineer named Marc Andreessen[4] who worked at the National Center for Supercomputing Applications at the University of Illinois, where he discovered the concept of the World Wide Web as defined by Tim Berners-Lee and Robert Cailliau, who worked at CERN in Geneva. Marc wrote the very first user-friendly web browser, called Mosaic. In 1993, Mosaic took the world by storm, because that was what people needed to explore the Internet: a user-friendly browser.

After graduating, Andreessen moved to California and met Jim Clark, who had founded Silicon Graphics, a company that built really cool workstations. Clark had fallen out with the current management of Silicon Graphics and had left the company. He was already wealthy, powerful, and extremely well connected in Silicon Valley.

The two hit it off, and Andreessen and Clark together founded Netscape Communications, based in Mountain View, right in the heart of Silicon Valley. Netscape built a commercial version of the Mosaic browser that Marc had built back in Illinois, and when it was launched, it took the world by storm. Andreessen was 23 years old at the time.

Jim Clark had the insight to capitalize on Mosaic's huge success and pushed Netscape toward an IPO on Nasdaq in 1995. It was one of the most memorable moments in Silicon Valley history: the stock was to be offered at $14 per share, but a last-minute decision was made to double the price of the initial offering to $28 per share. The stock's value soared to $75 during the first day of trading, making Andreessen an instant billionaire and landing him on the cover of *Time* magazine.

THE HALO EFFECT

Given the halo effect, many people thought Netscape's managers were the most talented group of people in the world, set on the clearest strategic course imaginable. Nothing could have been further from the truth. The company grew like wildfire to more than 2,500 employees, acquiring many small start-ups and interesting technologies, but the core question, "How will we make money?," still lacked a clear answer. Netscape had built a colossal loyal following by supplying the world with free Internet browsers, but where should it go next?

Netscape had all the potential to do a Schumpeter on probably the smartest software company in the world at that time: Microsoft. Microsoft had built its strategy entirely on software to run on your computer, and all of sudden companies like Netscape were saying, "Nah, you don't need that—you can do all that on the Internet." Microsoft had been very slow to realize the threat from the Internet, but when it woke up, it came out with all guns blazing. The resulting battle is known as the "battle of the browsers," pitting Microsoft's (free) Internet Explorer against the (almost free) Netscape. Netscape couldn't cope. It went into a tailspin, suffered from bad product releases, and lost both the battle and, ultimately, the war.

Netscape collapsed, eventually selling itself to America Online in a stock-swap deal that spelled the death of most of its activities. A sad ending for a company that had had the potential to conquer the world.

But when you ask many people in Silicon Valley which company they respect the most, they'll often say, "Netscape, because it was born, changed the world, and died in just 48 months."

And if you look at the fallout, you can clearly see the phoenix from the ashes. Netscape created an entire industry of Internet companies. Without the Netscape effect, there would be no Google, no Twitter, no Facebook. The rebirth wasn't just conceptual: the collapse of Netscape created opportunities for its 2,500 employees to put some of the wealth its IPO had created toward their own entrepreneurial dreams.

And let's not forget the boy who started it all. After the death of Netscape, Andreessen became one of the most influential venture capitalists in Silicon Valley. His firm, Andreessen Horowitz in Menlo Park, has fueled many next-generation Internet entrepreneurs.

WRITTEN IN THE STARS

As a kid, I was fascinated by stars. I learned that our own sun, our loyal golden disk in the sky, was actually just the same as all those stars we could see at night.

And the most fascinating element was that there were plenty of different types of stars out there, each evolving differently. Enter the fascinating world of brown dwarfs, white dwarfs, red giants, supernovas, and black holes. Sounds like a Tolkien band of brothers.

And the more I think about it, there is a great analogy between the evolution of companies and the evolution of stars. Just as no star will live forever, no company is truly built to last.

And, as with stars, their death can take many, many forms.

I have no intention of turning this book into a *Stellar Evolution for Dummies*, but bear with me.

Stars are typically born in what is known as a molecular cloud, a higher-density region of space that consists mostly of hydrogen, perhaps with some helium and a few heavier elements. It begins with a gravitational instability within such a molecular cloud. This can be triggered by shock waves from nearby supernovas (the massive explosions caused by the collapses of huge stars) or collisions between galaxies.

So, the death of one star can trigger the birth of a new one. This has Schumpeter written all over it.

As the molecular cloud collapses, individual pieces of dense dust and gas form what are known as globules. As these globules collapse and the density increases, the gravitational energy is converted into heat, the temperature rises, and a star is born.

There are small stars, medium-sized ones, and big ones, and each type has a completely different life.

Let's start with the small stars. Small stars are typically about 10 percent the weight of our own sun. They grow over time, fusing the hydrogen to produce helium and becoming hotter and higher in pressure. At the end of their lifetime, they shrink and become what are known as brown dwarfs. These are the quiet leftovers of small stars.

But stars that are more like our own sun have a completely different life. Our sun will become bigger and bigger over time, eventually growing to become a red giant that will be 250 times larger than the current sun, and swallowing most of our current solar system. That's still a couple of billion years from now. But in the end, the massive red giant will collapse, producing a white dwarf at the heart of a planetary nebula.

Yet the most spectacular life spans belong to the big stars. These stars have 10 times the mass of our sun, and grow rapidly to become what are known as supergiant stars. These enormous furnaces become huge extremely fast, but the result is catastrophic: when they collapse (and they all do) the result is a supernova—an explosion of enormous proportions.

And when those supernovae die down, the result is either a tiny, tiny neutron star that still emits radiation or a massive black hole—a tear in the fabric of space-time that sucks all energy, even light, into its black and deep abyss. This is the stuff that Stephen Hawking dreams about at night.

BUILDING STELLAR COMPANIES—AND BLACK HOLES

So what does this have to do with business? It gives us a way to understand the evolution of companies. In fact, this evolution can be mapped directly onto stellar evolution charts.

The molecular cloud is the entrepreneurial ecosystem that is the birthplace of the start-ups of this world. This nebulous crowd of talent, not hydrogen, is the fuel that start-ups are made of. And when an explosion happens—the collapse of an industry, or the fall of a mighty icon like Netscape—there is the chance that some of this talent will get together in enough critical mass to form a new entrepreneurial star.

And there are different types of start-ups. There are those that are the equivalent of small stars: great companies, with wonderful people and brilliant ideas, but just not big enough to really shake up the galaxy. Sure, these companies will grow, and they will do many useful things, but they won't put a dent in the universe. Many of them will fade away, become irrelevant, or be fodder for acquisition. They are the Getarounds of this world—a peer-to-peer car-sharing marketplace that was predicted to transform the transportation industry in a Forbes article entitled "10 Greatest Industry-Disrupting Startups of 2012."[5] Yet everybody is talking about Uber now, when it comes to disruption in this sector, and Getaround's success seems but a distant memory.

Then there are the sun-sized companies. They burn venture capital money fast and grow spectacularly, becoming leaders in their fields and kings of their markets. But when companies grow, they consume more and more energy, and they can implode under the weight of their own structures—become too slow to move, too fat to react, and too rigid to remain relevant. Look at the history of Polaroid: although it once ruled its market, it eventually folded and collapsed. You might say it became a white dwarf—the name Polaroid remains, and you can even buy film now for old Polaroids. But it remains a white dwarf.

But then there are the massive stars—the companies that receive the biggest halo effect. The ones we talk about in coffee shops and read about in magazines. The enormous companies that grow so fast that it is dizzying to watch. The companies such as Netscape that go from 2 people to 2,500 people in a matter of months, whose IPOs rock Nasdaq, but that burn fuel at an alarming rate, soon becoming so vast that when they implode, a supernova is unleashed.

Instead of creating a black hole, Netscape's supernova became a neutron star. Its radiation still shines as a beacon of hope and inspiration for the next generation of entrepreneurs. Netscape lives on.

The result of a supernova can also be a black hole. Enron is perhaps the most visible example in recent history. Named America's Most Innovative Company for six consecutive years by *Fortune* magazine, Enron grew from practically nothing to a staff of more than 20,000 in 2001. Its revenues reached nearly $101 billion during 2000. And then it went supernova. It turned out that the stellar growth was funded by nothing less than institutionalized, systematic, and creatively planned accounting fraud, and Enron collapsed, leaving behind a black hole so deep that an enormous number of people lost entire fortunes.

I don't believe in *built to last*. I think we should look to the stars and understand what makes them come into being, grow, evolve, collapse, and then reinvent themselves all over again.

I believe in creative destruction, in the phoenix from the ashes. I believe in Shiva, the god that is both destroyer and creator.

CHAPTER 8

STRATEGY FOR THE AGE OF NETWORKS

Companies will have to get in touch
with their inner innovation networks,
understand how to turn them into fluidity,
and avoid becoming rigid
corporate structures.

"Entropy: a vision of the ultimate, cosmic heat-death, a tonic of darkness and the final and ultimate absence of all motion."

—*Thomas Pynchon, from the short story "Entropy"*[1]

I did not choose this quote haphazardly.

Thomas Pynchon is perhaps one of the most influential writers in American literary history, one of the greatest and darkest minds to put his thoughts in writing, and yet no one has ever seen him. Well, almost no one. It's not uncommon for great writers to be reclusive, shying away from fans and hiding away in the comfort of anonymity. J. D. Salinger, one of the great fiction writers of the twentieth century, was a recluse. But compared to Pynchon, Salinger looks like Elton John.

Pynchon's most celebrated novel is his third, *Gravity's Rainbow*, published in 1973. It has often been compared in artistic value to James Joyce's *Ulysses*. Pynchon has a gift for combining the realms of culture, history, science, and technology with the absurdity of human behavior and the preposterousness of organizational insanity. You can understand why he is one of my favorite authors.[2]

But he hates public exposure. After the publication and success of *Gravity's Rainbow*, Pynchon was awarded the top prize at the National Book Awards ceremony in New York in 1974. A huge crowd had gathered to attend the ceremony, curious to see the by now famously reclusive author. They didn't know that Pynchon had actually sent a comedian to accept the award on his behalf. Since no one had ever seen the author, most guests assumed that it was Pynchon on stage. Soon after, articles started to appear in newspapers stating that Pynchon was in fact a pseudonym for J. D. Salinger. Pynchon's written response to this speculation was simple: "Not bad. Keep trying."[3]

Perhaps Pynchon's greatest literary obsession, running throughout almost all of his work, is the central theme of entropy—the ultimate "heat-death" of the universe, predicted by the merciless laws of thermodynamics. Eventually, over time, systems become affected by an ever-increasing prevalence of mediocrity, where everything becomes a constraining grayness, and eventually all motion and all life fade away.

FUN WITH CREEPING DEATH

Exactly this phenomenon seems to be affecting organizations as well. Companies seem to be battling the same homogenizing force today, moving so slowly and innovating so feebly that the ultimate end state for them seems to be "absence of all motion." Just as Pynchon is obsessed with entropy in general, I have become besotted with the concept of entropy in organizations.

Companies seem to be obsessed, as we mere mortals often are, with longevity—constantly pondering their corporate life span, contemplating their professional shelf life, and longing to become "built to last." But no one is. Corporate history reveals that organizations, just like us mortals, have one inevitable destiny: death.

We've addressed the VUCA model at length in this book, and the world of volatility, uncertainty, complexity, and ambiguity seems to be affecting nearly every market. It will be the playing field on which companies will need to find the fluid strategies they will need if they are to survive. In a VUCA world, the two essentials seem to be the parallel properties of speed and agility—to be fast, and to be nimble.

But is that enough? Can companies really use speed and agility to escape entropic death? And what makes organizations retain those capabilities? How can companies that grow remain capable of beating the odds of grinding to a halt?

The root cause of the entropic death of organizations is our inherent focus on the *structures* of organizations, rather than the intrinsic *dynamics* of these same organizations. We seem to revere the skeleton—the bony structures that grow but become fossilized. Instead, it is the flow of information that we should observe and cherish.

The core element enabling the survival of an organization, in my opinion, will be to rediscover the inner network—the core innovation network inside a company—and nurture that network in order to thaw the company's frozen capabilities. Many companies have lost touch with these hidden networks, which are tucked away deep in the burrows of organizations, hidden like a reclusive author.

It does not have to be that way, though. Barclays Bank, for instance, was able, in spite of its size and age, to spot these innovation networks through an internal start-up program that was expanded after the successful launch of its person-to-person mobile payment platform Pingit in 2012. The latter was introduced in just seven months, a process that would normally have taken three years. Barclays' program now has 40 business ideas in development.[4]

General Electric is a massive company with revenues of about $150 billion (£91 billion) and a market capitalization of nearly $265 billion. Yet still, it has integrated innovation into every part of its businesses. The company has thousands of researchers and engineers all over the world looking for new solutions. Its smartest move, however, was not just to depend on the intelligence within, but to experiment with open innovation. For instance, in order to respond even faster to the changes in its markets, GE invited external engineers to take part in design competitions.[5] A good example is its open innovation challenge targeting opportunities to reduce greenhouse gas emissions: looking for new uses for waste heat and improved efficiency of steam generation.[6]

Innovation is indeed about finding and jump-starting that inner core, but it is as important to connect and collaborate with those who have the ability to look beyond a company's limits and methods.

But the clock is still ticking for a lot of organizations. The advent of the age of networks has accelerated the rate of change of markets. And those markets are being redefined by the laws of the network, behaving differently as the lifeblood of information flow fuels the transition from markets to networks.

The core dynamic at play here is a simple rule: if the outside world starts to behave more and more like a network, you will have to start behaving like a network on the inside as well if you want to escape entropic death.

FROZEN, FLUID, RIGID, AND SUPERFLUID

This is where I believe we can originate the concept of *thermodynamics of organizations*. Why are some companies fluid (agile and capable of responding faster than the market), and why are some companies frozen, incapable of responding fast enough?

Let's explore this a little further. When we talk about *frozen* organizations, it has a negative connotation, but it doesn't mean that these organizations aren't moving anymore. On the contrary, many frozen organizations are incredibly busy. They are bustling with talented people who are executing on processes, serving customers, delivering products to the markets, and trying to make their operations more smooth, more reliable, more productive, and more effective. There's nothing wrong with that—at least, not if you're trying to optimize your current offering and position. But if you're trying to outsmart the competition, innovate to take advantage of new trends and opportunities, and be faster than the speed of the outside clock, you don't want to be frozen.

Fluid organizations, on the other hand, can change quickly, not only in their offerings, but also in their form. They can shape-shift their organizations with such flexibility that they can easily adapt to market trends and changing customer behavior. Centrica-owned British Gas, for instance, the largest U.K. energy company, which has about 10 million domestic customers, sensed a disruption in its industry, with newcomers like Nest offering smart meters. Rather than continuing to optimize its services and concerning itself with "pipes and wires," it decided on a start-up approach and introduced its very own smart-metering subsidiary, Hive. Not only did it recognize a shift in the market over time, but it bypassed its own complex and slow corporate structure with a lean start-up approach. The result was fast— and just in time—innovation, with Hive's product development taking days and weeks rather than months and years.[7]

The extreme of these fluid organizations would be what I call *superfluid* outfits. These are so liquid, and so incredibly nimble, that they can shape-shift overnight—radically change direction on a whim. Typically you see this in the quintessential Silicon Valley start-up. With hardly any corporate structure or bureaucracy, these young firms can turn on a dime during that crucial period when they are experimenting with radically new propositions or products.

PayPal was such an organization in its early years. Its founder, Max Levchin, once confessed that he initially envisioned PayPal as a cryptography company and only later as a means of transmitting money via PDAs. After several years of experimentation and failing fast, it installed itself successfully as an online payment system that is now used by millions of people.[8] The roots of Flickr lie in the development of an online role-playing game from gaming start-up Ludicorp. The founders luckily recognized the potential of simplifying photo sharing online, and their company was eventually purchased by Yahoo![9] Twitter started as a podcast delivery system; Intel sold computer memory, not microchips; Microsoft wanted to build software tools … I could go on here, but you catch my drift.

Superfluidity is actually a scientific term as well. It is one of the two states we've discovered so far of what we call *macroscopic quantum phenomena*— manifestations of the surreal world of quantum mechanics in our "normal" world. The other state is superconductivity. When a material is superconducting, you can run an electric current through it without any resistance.

This seems unearthly, but it's one of these materializations of the strange world of quantum mechanics in our normal lives. Well, maybe not entirely normal, since you have to cool ceramic materials down to almost absolute zero in order to get superconductivity (although scientists are closing in on materials that will have this superconductive quality at room temperature, which will completely change our use of electricity and magnetism, potentially propelling us into the world of the Jetsons, with electric cars and engines that run on virtually zero energy).

FLUIDITY AND START-UPS

Start-ups are superfluid—or they should be. Unfortunately, superfluidity often comes up against the other big driving force of nature in the realm of start-ups: venture capital.

Start-ups need two things: a healthy dose of optimistic, ambitious, and talented innovators, and a heap of cash to turn those innovators' dreams into reality. The venture capital industry and the start-up culture form a natural symbiosis. Where the two meet, amazing things can happen, and Silicon Valley is the epitome of that union of talent and money.

But the strangest thing happens when a start-up is funded by venture capital. Before that moment, when the embryonic start-up is still fueled by money from the famous trinity of friends, family, and fools, the company is superfluid to the extreme. But the moment venture capitalists put money into a company, their aim is to take a multiple of that money out. That's their business; that's how they survive. And that means that the venture capitalist wants the start-up to *focus*—to carry out the wonderful business plan on the basis of which the VC made the investment. To quit mucking about and just carry out the plan. In other words, to become less fluid. This can be extremely frustrating for the entrepreneurial founders of the company.

I have had the extreme pleasure—and luck—of founding three technology companies. The moment you launch a technology start-up, you're looking for money: money to grow; money to hire the right talent; money to establish markets; money to expand; money to innovate. But the moment you find people with serious money to invest in your company, you are confronted with your investors' desire that you focus, the most horrible word you can utter to a start-up. In the last company I started, the relationship between the founders and the investors deteriorated to the point where the company actually listed on the stock exchange in order to get rid of the conservative venture capitalists.

Most start-ups, as they grow and mature, naturally become less and less superfluid. But even a fluid company is amazing to observe. Fluid companies can still make sense of the signals that are coming from the market and respond to them. They can still look at technological or customer breakthroughs or trends and innovate with new offerings: new products or services. Fluid companies are capable of running faster than the market, which means that they can leverage the art of surprise. They might surprise customers with radically new offerings, shock competitors with market innovations that leave them gasping for air, or stun financial markets with colossal upsides in growth, earnings, or market share.

But when companies grow even further, they often start to freeze. When they focus intently on markets that they often dominate, they tend to emphasize the optimization of those markets rather than the innovation of new offerings. Frozen companies will look at becoming better at what they do, rather than at doing new things. They optimize operations, implement lean strategies, consolidate structures, streamline processes, and harmonize synergies. And to govern all that, they often build very elaborate structures: bureaucracies, committees, and supervising boards. A lot of consultants are hired, producing the most amazing PowerPoint presentations.

One of my favorite examples is that of Gerber's tragic missed opportunity. In 1974, the baby food and products specialist saw its growth stagnating, so it decided to innovate. It would tackle the market of processed adult food. It seemed like an excellent idea, seeing that Americans were spending more and more time at work and had less and less time to cook. Offering quick and wholesome foods to adults would certainly open up growth.

And it probably would have worked, had the company not been so focused upon operational efficiency. Instead of developing a new line of food, it basically tried to sell its existing baby products to adults, packaged with a different label and sold in a different aisle. Needless to say, this did not work—at all. The sad part is that the original intent was a good one—anticipating a real change in food habits—but the company was so focused on its processes and its margins that it completely botched things up.[10]

But the danger isn't freezing in itself, exactly. The danger is that these companies become so sluggish that they start to focus more on themselves than on the outside world. They lose touch with their customer base and become completely narcissistic and introverted, unaware of the dynamics and chatter of markets. And then they become *rigid*—a state of atrophy from which a company cannot escape. You may be able to turn a frozen company into a fluid one, but you can't turn a rigid company into anything. The only fate for a rigid company is entropic death. And soon.

THE THERMODYNAMIC CYCLE OF ORGANIZATIONS

One of the fundamental notions of thermodynamics is the notion of reversibility. Some processes in nature are reversible: they can be undone, or reversed. Some processes in nature are irreversible: they cannot be undone. Turning water into ice is reversible; you can melt the ice back into water. Turning sand into glass is irreversible; you can't turn a windshield back into a beach. The modern science of thermodynamics is actually quite recent, but it deals with fundamental processes that humans have observed in nature since their very existence: the transition of ice into water, water into vapor, and back. Of course, these processes depend on external factors such as temperature and pressure. Water boils at 100 degrees Celsius, at least at sea level, in most of the world (except in the United States, where it boils at 212 degrees Fahrenheit). But water will boil at only 80 degrees Celsius at the top of Mount Everest. The laws of thermodynamics govern the way the same chemical, H_2O, materializes into frozen ice, liquid water and nebulous vapor, under the conditions of heat and pressure.

Thermodynamics is very tangible and very practical. It has given rise to a ton of pragmatic applications that have made our lives better, including the air-conditioning systems that keep our bodies cool and the cycles of compression and expansion of Freon gas inside our refrigerators that keep our food cool.

I believe the time is right to apply the simple concept of thermodynamic cycles to the world of business organizations. I believe the fundamental transition of organizations from superfluid to fluid to frozen and (irreversibly) to rigid can be understood through a very simple organizational thermodynamic chart.

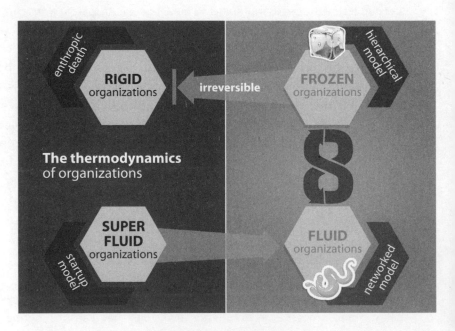

The companies that are capable of escaping entropic death will be those that can cycle between fluid and frozen states, like General Electric, Google, or Eli Lilly. Frozen isn't always bad. In fact, parts of an organization—perhaps even most of it—can be locked in to focus on a market mechanism that the organization understands really well. If core processes are optimized, this frozen mechanism can generate an amazing amount of profit and wealth. But if the organization lacks sufficient innovation liquidity, chances are that it will slowly and irreversibly become rigid. Like Blockbuster, which failed to react appropriately to the entrance of the likes of Netflix in its market. Like Kodak, which failed to recognize the disruptive power of digital photography. Or like Borders, which made the tragic mistake of underestimating what the Internet would mean for bookselling. The challenge for companies is to keep parts of the organization, parts of its skill base, in a fluid state, allowing room for experimentation, room for understanding the agility of the markets, and room to innovate fast enough to outsmart the market.

But how can a company create a dual state of frozen and fluid? How can a company be flexible enough to escape the irreversible entropic death trap of rigidity? For the answers, we will have to take a closer look at the brain.

THE TRIUNE BRAIN

It seems that nowadays no book or lecture is complete without a pseudo scientific exploration of the human brain in there somewhere. Oh, and it should include at least one visual diagram of the brain, pointing out the neocortex and perhaps some more exotic parts.

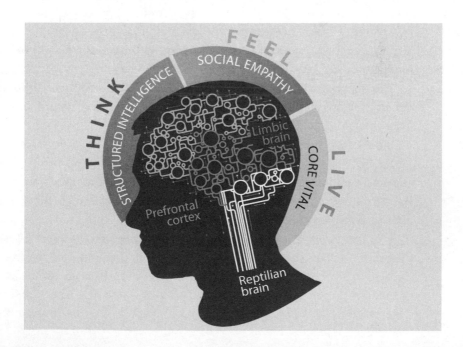

Actually, brain structure is extremely relevant to the discussion of fluid and frozen organizations.

The triune brain is a model proposed by the American physician and neuroscientist Paul D. MacLean. In his model, our brain consists of the reptilian complex, the limbic system (the paleomammalian complex), and the neocortex, also known as the neomammalian complex.[11]

Reduced to its essence, the reptilian brain keeps us alive. It's the core engine of survival.

The limbic system keeps us coherent as a species. The way we interact, survive in groups, and feel empathy are all part of the limbic system. It's the social side of the brain.

The third component, the neocortex, is the part that we're so proud of as human beings. Our capacity for logical thought, structured analysis, and the deep understanding of how pivot tables work in Microsoft Excel is all thanks to the marvels of the neocortex. It is the pinnacle of human evolution, and it's also slow. Incredibly slow.

The reptilian brain processes millions of bits of information per second, every second—all the inputs signaling motion, temperature, pressure, and other forces coming from the myriad sensors of your body. But you don't even notice it.

By contrast, the famous neocortex is slow as hell. This section can process only a handful of information. Some researchers claim that our illustrious neocortex can process less than 100 bits of information per second. In the same second, the reptilian brain has processed *millions* of bits. In terms of raw power, the reptilian brain blows the neocortex away. But the reptilian brain can't do pivot tables, so the neocortex gets all the publicity.

The neocortex is your conscious brain. This is the part that can think about itself and reflect on the conscious will that you have, the full control over your ideas, personality, behavior, and actions.

But recently, many researchers have started to raise big question marks about free will. A tremendous amount of research has gone into the way we perceive choice. As we saw in our discussion on the evolution from markets to networks, the field of neuromarketing relies heavily on being able to influence our subconscious.

Let's face it: our conscious brain is probably overrated. If I throw a baseball at you, at high speed and without prior warning, chances are that you'll be able to catch the ball (or at least duck) before it strikes you in the face. Hopefully. But you didn't have time to think it over, or to weigh your options. Your response was processed by the reptile within—not by the vaunted conscious part of your brain.

And your conscious brain hates that. There's nothing it despises more than the idea that it doesn't have everything under absolute control. So it makes up a story that it caught the ball on purpose, by sheer force of logic and reasoning. But that's a lie. Recent neurostudies have shown that the logical and structured part of our brain is constantly making up stories to assure itself that everything is under control. But in reality, many of our actions and behavior are completely out of the scope of our free will, and are instead driven by our unconscious.

The theory of the triune brain has been contested by many researchers, who claim that it is too simplistic. But I like the simplicity of there being three big factors that make up the workings of our gray matter—survival, social, and logical—and I think this theory works very well as a model for understanding our behavior.

THE TRIUNE NETWORKS

Just as our gray matter is made up of three parts, I think our organizations have three distinct internal networks that together form the triune networks of companies:

1. The core innovation network
2. The social network
3. The structured network, or hierarchy

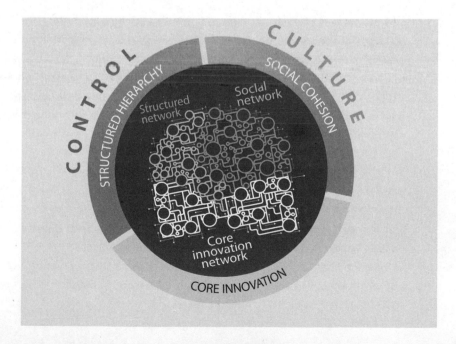

The core innovation network is the essential network of people that make innovation happen at your organization. When a company is small, this core survival network is clearly visible, but as a company grows, it becomes reclusive and is no longer clearly visible. But it is vital. It is the core group of people that you would need if you ever had to start over.

It's the nucleus that could go across the street and start a competitor to your business. It's the group that might find the next idea to take your business to a new level—and the group of people you'd be most horrified to see your competitors get hold of. But most companies have lost sight of who participates in their core innovation network. It's like a secret society buried deep in the dungeons of the corporate hierarchy.

A company is also a social network—a group of people who work together, relate to one another, and spend a lot of time with one another. Companies are filled with people who become friends (or enemies, for that matter) and develop relationships and ties. These social networks are crucial in developing or changing a company's culture, and they have become vital in understanding why people want to belong to a group or engage with an organization. In the very same way that customers relate to the image of a product, these cultural networks inside companies are vital conduits for the emotional sentiment that runs through an organization. It's no wonder that the companies that attract the smartest and most talented people— like Google, SAS, or salesforce.com—invest a lot in keeping their personnel happy, inducing a tight community feeling, and having their staff function like a "tribe."

The final part of the triune network tends to be the most visible: the hierarchical network, often depicted in the infamous org chart. This network determines who does what and who reports to whom. The hierarchical network has a huge role to play in a company, especially in areas where a military command-and-control structure is needed to maximize operational efficiency. Unfortunately, this third network is often the only one that really captures the attention of top management.

WHY START-UPS ARE MAGIC

As I've mentioned before, I've had the enormous privilege and pleasure of being involved in three technology start-ups, early on in life.

Few activities are more intoxicating than doing a start-up venture. With a small group of like-minded enthusiasts, you're out to break the rules and cross frontiers. You're out to change the world.

All three of those start-ups were exhilarating and fascinating. But each one was also an enormous risk, a huge effort, and a tremendous fight. None of them was easy.

One of the things I've observed in these start-ups, and in many other start-up activities I've observed, is that when you start out, the three networks we just discussed are completely in sync. As a matter of fact, they're indistinguishable. The triune networks sync completely.

Your core innovation network *is* your start-up; these are the people that you believe can change the world, the ones who are willing to take huge personal risks and are excited about the thrill of being part of something magical. But they're often also your social network. Very often, the people you hire at the beginning of a start-up are your friends, the people you met in your dorm, friends of friends, or the people you're comfortable with. And since in a start-up, you spend pretty much all your waking hours on your project, you're likely to have no other friends.

But the start-up is also the structured network, the hierarchy. People who believe that there is no hierarchy in a start-up have obviously never really experienced a start-up. Hierarchies in a start-up are fierce, clear, and unambiguous. They're undocumented, but they're fully understood. Everyone in the start-up knows exactly who calls the shots and who gets the credit. All the developers know who's the top coder, who controls the architecture, and who decides what to build. All the salespeople know who the top dogs are; the market guys know who makes the calls.

The structured network is the same as the social network is the same as the core innovation network. The three networks map perfectly. As a start-up grows and matures, however, the three networks tend to grow apart. They become less intertwined, sometimes separating entirely.

The most magical moment in the life of any start-up is when, for whatever reason, you post your first org chart. It could be that you've just landed a big round of financing or venture capital funding, and the venture capitalist on your board would like to understand how this joint is organized: who's in charge, and who reports to whom.

When the first org chart is posted on the wall, it can be a hilarious moment. 'Really? George is the lead architect reporting to the CTO?' Everyone in the company knows that George doesn't know shit about architecture! The guy who really knows the ins and outs of the product architecture is José, who lives in the basement. Literally. He bunks out on an inflatable mattress next to the development servers, eating pizza by the blinking lights of his machines. José built the architecture, lives the architecture, *is* the architecture. But you can't show José to investors.

He probably doesn't even own a suit or a pair of presentable shoes. In contrast, George wears an Italian suit and shiny wingtips—and he can draw really cool diagrams in PowerPoint that are very convincing at board meetings.

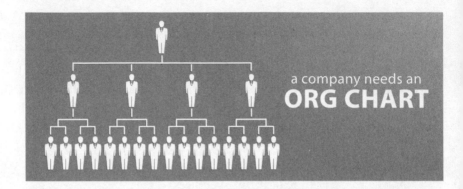

a company needs an
ORG CHART

So George is the lead architect. George is happy—he's on the org chart. But José is happy, too. He's happy sleeping with his servers and building the next level of the product's architecture. He's also very happy that he doesn't have to present or report to the board, because the idea of that terrifies him, and he'd probably have to buy a suit, which he hates. George is happy. José is happy. The board members are happy. Everyone is happy.

But as the start-up grows, it loses touch with its innovation network. George is promoted. He rises through the ranks as more and more prestigious jobs appear on the org chart that require the essential skills of devising beautiful PowerPoint slides and uttering meaningful sentences about flexible design strategy and modular scalable components, while wearing impeccable suits. Eventually, George becomes CTO.

But unlike in the beginning, when everyone knew that José was the real genius behind the product's architecture and the core innovator in the organization, now few people have any clue what that weirdo with the beard and the flip-flops and the "Yoda Lives Forever" T-shirt actually does down there in the basement. José never made it into the structured network. He dropped off the social network pretty quickly as well. But José is still a pivotal node in the core innovation network of the company.

And he's the first person I'd hire if I wanted to start a new start-up that would blow the old company out of the water.

INDUSTRY DISRUPTION

Start-ups are a petri dish on steroids when it comes to organizational evolution. The transition from an idea on the back of a cocktail napkin to a super-fluid organization is sheer magic. But it happens every day. There are more than 25,000 start-ups in the San Francisco and Silicon Valley area, all aiming to blow the previous generation of miracle tech companies out of the water. All aiming to become the next Google, Facebook, or Twitter.

Mark Zawacki is based in Palo Alto, at the epicenter of entrepreneurial disruption. Mark runs a company called 650 Labs, whose name is based on the telephone area code for Silicon Valley. He's a strategy consultant and angel investor who spends a great deal of his time helping traditional companies understand the magic of entrepreneurial innovation that is the hallmark of Silicon Valley—and helping them rub some of that magic onto their own organizations.

According to Mark, the biggest shift of the last couple of years is that Silicon Valley is moving from being the high-tech capital of the world to becoming the industry-disruption capital of the world. We all know that Silicon Valley gave birth to technology giants such as Hewlett-Packard, Intel, Cisco, Sun Microsystems, and Oracle. These companies all grew up in the intoxicating entrepreneurial landscape of Silicon Valley, where such tech giants could flourish. But that is exactly what they were: technology companies. According to Zawacki, today, the Valley and San Francisco are filled with start-ups that *use* technology; they employ the same technology geniuses from Stanford and MIT, but they are out to completely disrupt other industries.

Facebook is not a technology company, but it has completely redrawn the business model of advertising for the age of networks (University Avenue in Palo Alto has replaced Madison Avenue in New York as the epicenter of advertising innovation). Netflix is not a technology company, but it is redefining the foundations of the media industry, as well as the model of the television industry. Tesla is not a technology company, but it is completely disrupting the age-old model of the car industry. I could go on and on with many of the area's 25,000 start-ups.

When you look up synonyms for *organization*, you get equivalents such as *firm, corporation,* and *institution.* Those words say a lot. *Firm* connotes something fixed, solid, and unyielding. And I shudder whenever I hear the word *corporation*, which suggests something stringent, bureaucratic, and inflexible. The shadow of entropic death flutters nearby.

And when you then see the rate of innovation that these new innovative start-ups are bringing, and how much disruption they can bring to age-old industries, you can begin to understand what keeps companies from being swallowed by that shadow. Companies will have to get in touch with their inner innovation networks, understand how to turn them into fluidity, and avoid becoming rigid corporate structures that will get run over like deer staring frozenly into the headlights of the oncoming cars.

IT TAKES A NETWORK TO FIGHT A NETWORK

U.S. Army General Stanley McChrystal has had a long military relationship with the Middle East. He saw action in operations Desert Shield and Desert Storm, was the commanding general in Afghanistan in 2009 and 2010, and before that was the commander of the Joint Special Operations Command from 2003 until 2008. In 2010, McChrystal was forced to resign from his post after *Rolling Stone* magazine published an article called "The Runaway General." The politicians back home did not appreciate the candor, frankness, and criticism expressed in the article. The general had to take the fall.[12]

Today, Stanley McChrystal is helping companies understand why networks are important, and why old structures don't work any longer. To do so, he founded a consulting firm called McChrystal Group.

The heart of his philosophy is based on the things he learned while fighting in Iraq and Afghanistan, and his failure when he tried to use old military structures in a war on terrorism, where his adversary was not a structure, but a network.

The U.S. military was used to fighting structures similar to its own. Wars were mostly waged between centralized, hierarchical structures with military-style control, command, and discipline. That approach worked well in the two world wars and even during the cold war. But it failed in fighting enemies such as terrorist networks.

McChrystal learned this the hard way. In his own words, "In bitter, bloody fights in both Afghanistan and Iraq, it became clear to me and to many others that to defeat a networked enemy we had to become a network ourselves."

This became the central theme of his doctrine: "It takes a network to defeat a network."

As McChrystal describes it, in Iraq, Al-Qaeda operatives did not wait for memos from their superiors, much less direct orders from bin Laden.

Decisions were not centralized, but were made quickly and communicated laterally across the organization. He portrayed the fight with his enemy as a deadly dance with a constantly changing, often unrecognizable structure.

McChrystal tried to turn his old-style military structure into a network to match the workings of his enemy: "We had to figure out a way to retain our traditional capabilities of professionalism and technology, while achieving levels of knowledge, speed, precision, and unity of effort that only a network could provide."

The *Rolling Stone* article came too soon. McChrystal's resignation cost him the opportunity to try to transform U.S. military operations into full-fledged networks that would be better suited to this new kind of enemy. He may not be the first military commander in history to clearly see the strategic need to transform, but he's the first to define that need so clearly for the present age.

THE GOLDEN RULE

As we described earlier, the fundamental shift that we are now seeing is that markets are becoming networks. Customers form the heart of networks of intelligence, and old marketing concepts don't work any longer. We have to learn how to connect to consumers, influence networks of information, and understand the dynamics of network behavior.

These imperatives also have huge consequences for the way we see the evolution of organizations. The core dynamic at play here is a simple rule: as the outside world starts to behave more and more like a network, the inside of your company will have to start behaving like a network.

I believe that this is the single biggest challenge facing corporations today. When I wrote *The New Normal*, I was convinced that the rise of digital, and the fact that technology was rapidly becoming normal, was a wake-up call for companies, alerting them to a need to pay attention to an age in which they would have to communicate with customers in a different way as a result of the abundance of digital communications.

But now I think that digital was only the appetizer. The New Normal is just the foundation for the age of networks, a much more fundamental transition in business strategic thinking, in which companies will be confronted with the interplay of network behavior throughout their businesses, markets and workforce.

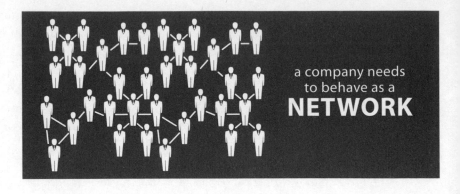

a company needs
to behave as a
NETWORK

THE HOLY TRINITY IN THE AGE OF NETWORKS

In the world of business, the holy trinity is the combination of a company's business strategy (*B*), the company's market (*M*), and the company's workforce (*W*). The correlation among these *BMW* elements defines the way a company operates, grows, and evolves.

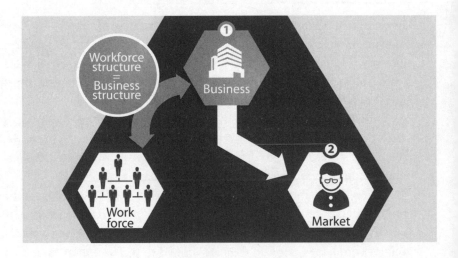

In the Old Normal, a company typically drove its business strategy forward with an inside-out approach. It would devise a strategy for conquering or cornering a market, deduce what kind of offering it would need in order to make that happen, and then project that offering onto the market, figuring out what kinds of customer segments to target with its offering. The structure that was applied to the workforce often mirrored the business structure of the organization. Employees would be drafted into departments and silos that imitated the business units and departments of the company.

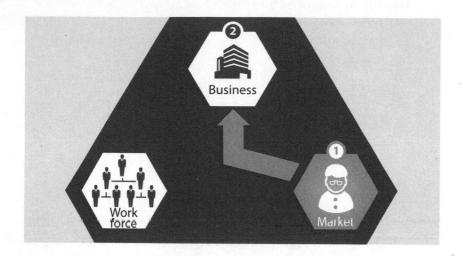

With the advent of the age of networks, that model fundamentally changes. As the consumer becomes more connected and empowered than ever before, it flips. When markets become networks, companies will have to understand how to tailor their business for these newly empowered customers—and learn to mirror the form that empowers them: the network.

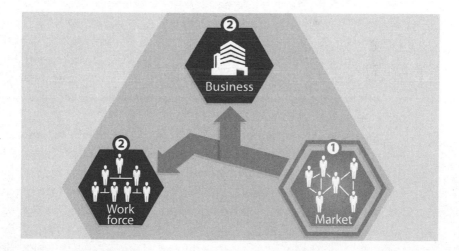

This will have a profound impact on the way we organize our workforce as well. When the golden rule of "It takes a network to serve a network" is applied, this means that our old bureaucratic structures will no longer suffice. These structures won't be dismantled entirely, since they will still function in frozen parts of the organization. But in those areas where markets are heating up and becoming networks, the frozen organizational structures of the past are no longer valid. In these networks, where fluidity is the norm, organizations that are hoping to innovate will have to become networks as well.

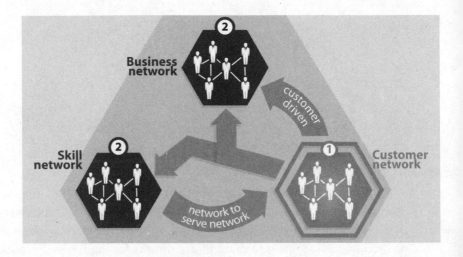

The full picture becomes visible when we attempt to label these transitions. Markets are turning into networks of information, serving *customer networks*. The workforce of an organization has to understand the implications and limitations of the structured hierarchies, and adopt the notion of *skills networks* in order to survive. And eventually, network-based thinking will allow a company to dramatically rethink its role, decompose its core offering, and work differently with partners and suppliers, eventually becoming a *business network* itself.

Companies are no longer smart enough, fast enough, or innovative enough to survive on their own. And a lot of larger corporations have started to understand the value of networked collaboration. It's like Toyota Motor Corp. partnering with Microsoft to develop a software platform for managing the information systems of electric vehicles. Or like the Coca-Cola Company working with Heinz to produce a bottle made of 100 percent plant-derived material. Or American Express partnering with Foursquare to offer location-based discounts from participating merchants.[13] Or AstraZeneca offering easy access to its library of clinical compounds on a collaboration platform to facilitate research.[14]

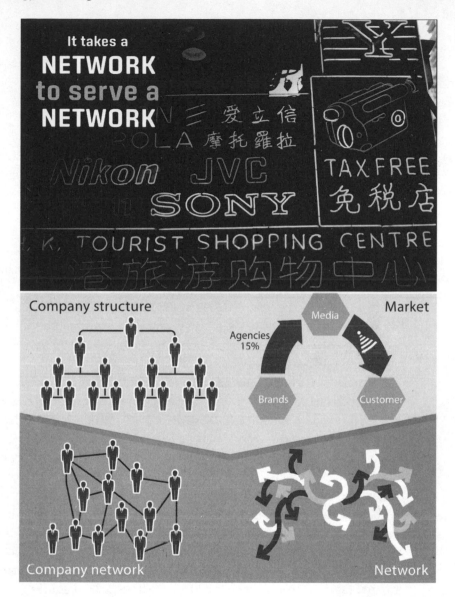

Philips even has numerous collaborations with companies from different industries. It has, for instance, innovated together with Nivea on a conditioner-dispensing shaver, with BASF on a solar car roof, and with Aerogen on a noninvasive ventilation system for patients with breathing problems.

Another great example is that of the Solar Impulse partner team of—among others—Omega, Schindler, and Deutsche Bank that came together in 2004 and is led by Solvay. Together they built the very first aircraft in history to fly both day and night without using fossil fuels.[15]

WAVES OF DISRUPTION

In the coming years, with the full force of VUCA hitting the world of business, we will see waves of enterprise business model disruption that reshape entire markets and market systems. When the laws of networks behavior overturn our understanding of market dynamics, we will see that the companies that survive will be those that have understood the need to drastically rethink their organizational fabric.

Whenever radical shifts of theory occur, whether in science, culture, or philosophy, the prevailing inclination is to discredit this new way of thinking, precisely because it is so radically different from the past. That has caused great frustration for the creative thinkers behind the novel ideas. Take Ludwig Boltzmann, one of the greatest scientists to come out of Austria. Born in 1844, he was a physicist and philosopher who was one of the key thinkers in the radically new field of thermodynamics.

Boltzmann was a genius, succeeding his teacher Joseph Stefan as professor of theoretical physics at the University of Vienna, where he taught physics and lectured on philosophy. Boltzmann was a great teacher, whose lectures on natural philosophy received a considerable amount of attention. His first lecture was so successful that, although the largest lecture hall at the university had been chosen for it, people had to stand all the way down the staircase. Word of his genius led the emperor of Austria to invite him for a session at the palace.

Boltzmann's peers in the field of science were not so enthusiastic, however. He was one of the key architects of the famous second law of thermodynamics, the entropy law, contributing insights from his background in statistical mechanics. His ideas were a radical departure from the prevailing scientific ideas at the time, and his critics were fierce in their attacks.

In fact, Boltzmann decided to become a philosopher in order to refute the philosophical objections to his physics. But he soon became discouraged by the ongoing criticism of his peers. On September 5, 1906, while on a summer vacation with his wife and daughter in Duino, near Trieste, Boltzmann hanged himself during an attack of depression. He was buried in the Viennese Zentralfriedhof, and his tombstone bears the inscription

$$S = k \cdot \log W$$

where S is a representation for the concept of entropy.

THE AGE OF NETWORKS

More than one hundred years after Boltzmann's death, the concept of entropy continues to fascinate mankind.

We have come to clearly understand the role that entropy plays in science, but I believe that entropy also plays a huge role in the way we have developed organizations and looked at markets. When companies grow, their internal capacity to innovate, and therefore survive, seems to diminish. Many large corporations seem to be heading for entropic death.

When markets become networks, new laws are at play. New theories will have to be developed, new concepts to be explored. Again, business can learn a lot from how science has looked at these new varieties of systems. Ilya Prigogine was the founder of the science of self-organizing systems. He was able to link his work in chemistry to the domains of biology and sociology. His work is seen by many as a bridge between natural sciences and social sciences. It is just those kinds of bridges that can be most helpful when adapting to a radically new reality.

Today we are about to see similar concepts in the world of business. When markets become networks, companies will have to become networks—and, in particular, scale-free networks. Unlike top-down hierarchies, these scale-free networks emerge from bottom-up interactions, and appear to be limitless in size.

In the age of networks, we have to understand that there is more to running a company than a top-down bureaucracy. In the age of networks, we have to understand the thermodynamics of organizations.

EPILOGUE

The biggest challenge for companies
won't come from the outside world. It will be to
change the dynamics inside their organizations in
order to match the speed of flow on the outside.
Unfreezing age-old hierarchical command-and-
control mechanisms won't be easy.

———

"The corporation will survive—but not as we know it. Organizations are critically important as organizers, not as employers. Often the most productive and profitable way to organize is to disintegrate and partner."

—Peter F. Drucker

SWARMS OF HARVESTERS

I have had the privilege of teaching at London Business School since 2008. I really enjoy spending time at the school, although I am not an academic. Far from it. When I graduated from my university in the field of computer science, I was last in my class. The kids who were at the top of my class all went on to have brilliant careers in academia. I had no choice but to start my own business.

I'm an entrepreneur, first and foremost, having spent the better part of 15 years nurturing and growing three technology start-ups. As an entrepreneur, you see them grow, fall, stumble, and rise up again, and, just like a parent, you're immensely proud when they mature and find a great new future. They get acquired, or they reach a bigger audience as a listed company. The thrill of listing my own start-up, being able to share the excitement of my company with the world, has been one of the most electrifying moments of my life.

That experience sets me apart from thoroughbred academics. It is perhaps also the reason why London Business School tolerates me as part of its faculty.

My favorite form of involvement with the school is the Senior Executive Program, which allows me to engage with senior business leaders from around the globe who travel to London for an intense four weeks of intellectual stimulation and strategic soul-searching. They get to spar with gurus like Rob Goffee on leadership and Costas Markides on strategy. And with me on the subject of technology and innovation.

A few years ago, during a lecture, I recounted a recent trip that I'd made to Silicon Valley with a group of insurance executives. We took them on a whirlwind tour, showing them the hottest technology companies and start-ups, and we spent a full day at Google. There we had the chance to expose the group to Google X, the research and innovation lab at Google, which had just unveiled the driverless car.

The insurance executives were dumbstruck. It is fiercely intimidating and strangely exciting to ride in a car that drives itself out of the Googleplex campus and heads out to San Francisco without a human behind the wheel, and only the algorithms of Sebastian Thrun to keep you safe. The executives were buzzing—and getting a little pale as well, since their company made a lot of money on car insurance. I imagined them thinking, "How on earth are we going to insure a driverless car?" But the most interesting question was: "Why did you guys do this? You're a search engine, for crying out loud! Why didn't Mercedes or BMW start this research?" Unfazed, the Google X people snorted back: "Nah. They're from the mechanical age. They think in oil, pistons, and gears. We're from the information age. We saw it as an information challenge."

I related this story to an audience that included a senior executive from a large manufacturer of agricultural equipment. He had seemed uninterested in technology, because he didn't think the world of digital would profoundly affect his business. His attitude was, "I'm not going to sell more combine harvesters because of Twitter." And he was right. But when I told the story of the Google driverless car, he became intensely interested.

A year later, I ran into him again at an alumni reunion. He came up to me and spoke passionately about the transition in his company. After the discussion on driverless cars, he started to realize that they could transform his business.

What if harvesters could drive themselves, 24/7, without farmers? The technology inside the vehicles is quite state-of-the-art, and it wouldn't be too difficult for his engineers to make the machines communicate and work together. (And a cornfield is much simpler to navigate than commuter traffic on the Golden Gate Bridge.) Soon he had trials up and running of autonomous tractors, transport carts, trailers, and combine harvesters capable of harvesting entire fields of crops without any human intervention.

That's when he got really excited. "Instead of selling these machines to the farmers, we could operate this as a service, with swarms of harvesters going from farm to farm to harvest the crops." His company had been selling machines since it was founded. Now he realized that the future might be to provide a service to the farming community, moving from region to region to follow the seasons.

Just as a footnote to the Google X people stating that car companies were not from the information age: there are indeed some "old-school" car companies that have understood the disruptive power of data. Ford, for instance, is collaborating with the insurance company State Farm in the development of an onboard system to analyze people's driving habits. The aim of the partnership is to calculate insurance costs in a much smarter and more tailored manner.[1] In other words, to use data so that the company can offer better services to its customers.

THE NETWORK OF THINGS

The earliest impact of the ideas behind *The Network Always Wins* might lie not in the domain of networks of human beings, but in networks of things.

For some time now, we've started to talk about an "Internet of Things," in which more and more connected devices are communicating with one another. Cisco estimates that by 2020, more than 50 billion connected devices will be alive and blinking on this planet. That's more than 10 devices for every human. The cost of sensor technology has been falling steadily; the cost of chips has dropped like a brick. We now have reliable networks in place that can be used not only to transfer e-mails, Facebook updates, or Snapchat selfies from humans to humans, but also to let things talk to each other.

And it's not just things. Cows, too. Never considered the most intelligent animals on the planet, they're rapidly becoming the most connected. Dairy-intensive countries such as Denmark have outfitted cows with sensors and network capabilities so that farmers can follow their dietary intake to optimize milk production. Networked cows are the most effective milk generators on the planet.

Meanwhile, companies like Philips, which is active in industries ranging from healthcare to lighting, are making all of their products network-aware. Lampposts have become active nodes in a network that can monitor traffic, violence, or accidents. Soon every light in your house will have an IP address and be part of a network, allowing you to control your lighting using your smartphone. Philips also recently introduced a connected toothbrush with built-in chemical sensors that can connect to medical databases via your phone.

Your house will be filled with things that are monitoring, connecting, and exchanging information. Your appliances will listen to the electricity network and know when it's most economical to turn themselves on. Your fridge will turn itself off when it thinks you're not home for a longer period. Your washing machine will turn itself on during the night to run a laundry cycle—after a fair bit of haggling between your connected machine and the electricity provider.

Your car will provide real-time data to the insurance company. We'll see the age of real-time, usage-based insurance, in which cars or bikes will be insured only when you're using them. Every time you drive your car out of the garage, fierce bargaining will take place between your vehicle and the insurance companies to give you the best possible deal on your ride to the supermarket. And you won't even know it took place.

This is the reality of the age of networks, where everything is connected. We will live in networked houses, in smart cities, in connected regions.

And that seems pretty scary to many people, because we don't trust machines.

When networks break down, they can cause enormous damage. The power blackout of 1965 in the northeastern United States was the worst in history, because of the connections shared in the power grid. An overload at one point caused a cascading failure throughout the grid. More than 30 million people in the greater New York area were left without electricity for up to 13 hours.

We are increasingly at the mercy of machines, in the hands of algorithms, and in the hands of networks of things that think.

It is widely known that airplanes can land and take off more reliably than any human pilot would ever be able to do with a few tragic exceptions. Driverless cars are more efficient, more economical, and safer than any human driver. So is it so bad to be at the mercy of a network of intelligent things?

Kevin Kelly is a great technological philosopher. He was one of the founders of *Wired* magazine back in 1993, when it became the central reference point for all things digital. Kevin is no longer the editor in chief of *Wired*, but he is the one who provided us with a very scary answer to the deceptively simple question, "What is technology?" Most of us would say that it's the collection of inventions, innovations, and creations that drive us further on the journey to progress as human beings. Kelly challenges that notion: "What if Technology is just a Vector, and we humans are mere servants to take the progress of Technology further?" According to Kelly, we are merely temporary helping hands in the evolution of technology. Humans like to think that we invented technology, but in the mind of Kevin Kelly, "Technology is using us to take itself to the next level." Technology is a vector, a force of nature, not the brainchild of the human race.[2]

Douglas Adams would have loved Kelly's view. In his own literary legacy, Adams depicts planet Earth as the "ultimate computer," and we humans as simply part of the fabric of this huge organic computer that will compute the answer to the "Ultimate Question of Life, the Universe, and Everything."

I'm sure we will see growing numbers of dystopian futures imagined by science fiction writers and Hollywood blockbusters focusing on everything that can go wrong when we enter the age of networked things. I can't even begin to imagine what will happen when artificial intelligence meets the world of networking and humans have no idea what kind of chatter is taking place on our networks—when intelligent lampposts start conversing with driverless cars and communicating with swarms of harvesters passing by fields of highly interconnected cows. Douglas Adams could have written a perfect nonfiction follow-up to *The Hitchhiker's Guide to the Galaxy*.[3]

FLOWING INTO THE FUTURE

Once we leave the New Normal and enter the age of networks, the fundamental change will be that information becomes a flow. We have to figure out how to make sense of that flow. That, in my opinion, will be the biggest challenge for organizations: understanding patterns and building insight into information that is constantly changing, relentlessly flowing, and continuously cavorting on countless networks.

Flow will become the essence of everything that determines the success of your business: flow in customer behavior, innovation flow inside organizations, and flow of value, knowledge, and meaning.

We're already seeing evidence of markets becoming networks. The fluidity in the marketplace, the speed at which markets move, and the relentless flow of information that is fueling the transition from markets into networks is becoming more visible each day. Companies are already being confronted with the transition from classic marketing approaches to learning how to influence networks.

But the biggest challenge for companies won't come from the outside world. It will be to change the dynamics inside their organizations in order to match the speed of flow on the outside. Unfreezing age-old hierarchical command-and-control mechanisms won't be easy.

Gary Hamel is another teacher at the Senior Executive Program at London Business School. I value his thinking on this subject. He says that most organizations have become overcontrolled bureaucracies, and that in order to thrive amid today's turbulence, organizations must become:[4]

- *Less* organized
- *Less* managed

- *Less* structured
- *Less* hierarchical
- *Less* routinized

This change in organizational fabric will be vital for survival. And according to Hamel, the change will be multifaceted, relentless, shocking, and revolutionary.

He makes a plea that we have to tear down old organizational structures and replace them with communities of passion, fueled by the flow of information.

It will take true leaders to turn companies back into fluid organizations. It will require vision and courage, and only genuine innovators will be able to lead their companies into the age of networks.

We don't fully understand the world of networks yet, but we do know its contours. For one thing, it is clear that the era of Industrial Age management techniques has come to an end.

John Kotter, one of the gurus on change management at Harvard, together with his colleague Ken Perlman, recently discussed this in a *Forbes* article about the advent of the era of the knowledgeable networker.[5]

They define this new breed as employees who are integrative thinkers with broad interests and connections. These networkers see how puzzle pieces fit together without needing to know everything about each piece. Instead, they know a lot of people and have a lot of information sources.

According to Kotter and Perlman, an organization will have to become a network of networks: people who stay connected and reach out for knowledge when it is needed. Organizations full of knowledge workers will continue to exist, but how they operate will change. Some organizations will fail because workers are not networked internally to share their developing knowledge.

This is vital for any company, even for the mother of all fluid companies. Google has to work hard to keep the fluidity inside its organization alive. In the last couple of years, Google has hired social scientists to study its organizational evolution. Collectively known as PiLab (short for People & Innovation Lab), these scientists run experiments in an effort to understand the best way to manage the firm as it explodes in size.

Among the early findings of PiLab is that middle managers actually matter, which proved wrong the presumption of founders Larry Page and Sergey Brin that you can run a company without anyone being the boss of anyone else.

My point is not that you should abandon all structures and replace everything with pure networked chaos. My belief is that when companies understand the different roles that the triune networks inside their organization play, they can ensure that the innovation network, the social network, and the structured network can all thrive and add to the organization's strategic evolution. The first thing most large companies will have to do is to rediscover their hidden innovation network and unlock its potential.

Many companies are beginning to realize that their Industrial Age structures, mechanisms, and strategies aren't helping them win the hearts and minds of networked workers and customers. In other words, there is a great disturbance in the Force. Perhaps the best thing we can do is listen very carefully to the next generation. That's why I dedicated this book to my children, who have taught me to look at the world of networks through their eyes.

This next generation has *networked* built in. Its members have grown up on Wikipedia and YouTube, and have learned to thrive in the networks of Facebook, Tumblr, and Instagram. They have ultimate mastery of Control-C and Control-V, and they suffer from continuous partial attention syndrome. But most important, they have learned that the more they put into the network, the more valuable they become, and the more valuable the network becomes for them.

For anyone in a large corporation, in an Industrial Age construct, I wish you all the best in unfreezing your company, your structures, and, perhaps most important, your own thinking.

Remember, where we're going, the network always wins.

ENDNOTES

PREFACE

1. Officially named the Dwight D. Eisenhower National System of Interstate and Defense Highways, after the president who signed it into law in 1956, this system spans the entire United States and is made up of highways that travel from north to south (the uneven numbers) and east to west (the even numbers). The system took 35 years to complete and cost hundreds of billions of dollars; in effect, it was the "costliest public works since the construction of the pyramids." It spans more than 47,000 miles (75,000 kilometers).

    In the hit comedy show *The Big Bang Theory*, Sheldon reprograms a GPS to spit out trivia questions, such as, "Which four state capitals are not served by the Interstate Highway System?" (Actually, there are six: Anchorage, Alaska; Carson City, Nevada; Dover, Delaware; Jefferson City, Missouri; Pierre, South Dakota; and Honolulu, Hawaii).

2. Negroponte's 1995 bestseller *Being Digital* predicted that the interactive world, the entertainment world, and the information world would eventually merge.

LET'S GET STARTED

1. Carlota Perez, *Technological Revolutions and Financial Capital: The Dynamics of Bubbles and Golden Ages*, Elgar, 2002.
2. Peter Hinssen, *The New Normal: Explore the Limits of the Digital World* (Ghent, Belgium: Mach Media NV, 2010).

CHAPTER 1

THE AGE OF UNCERTAINTY

1. Alex Abella, *Soldiers of Reason: The Rand Corporation and the Rise of the American Empire* (Boston: Mariner Books, 2009).
2. Herman Kahn and Irwin Mann, *Techniques of Systems Analysis* (Santa Monica, CA: RAND Corporation, 1957).
3. Herman Kahn, *On Thermonuclear War* (Princeton, NJ: Princeton University Press, 1960); Carl von Clausewitz, *On War*, trans. James John Graham (London: N. Trübner, 1873).

4. Moore's Law is the paradigm that states that the number of transistors on an integrated circuit doubles approximately every two years. This was first observed and stated by Intel cofounder Gordon E. Moore, who described the trend in 1965. The shortening of that period to 18 months is attributable to Intel executive David House.

5. Herman Kahn and Anthony J. Wiener, *The Year 2000: A Framework for Speculation on the Next Thirty-Three Years* (Croton-on-Hudson: Hudson Institute, 1967).

6. Herman Kahn would have hated the VUCA world that we now live in. But he saw it coming. He understood that the world of imponderables was a vital part of real life, outside the sterile models running on his computers.

7. Josh Sanburn, "10 Companies That Radically Transformed Their Businesses," *Time*, October 19, 2011.

8. Soren Kaplan, "Leading Disruptive Innovation," *Ivey Business Journal*, July-August 2012.

9. http://thefailcon.com/.

10. http://www.forbes.com/sites/hbsworkingknowledge/2013/02/25/lean-startup-strategy-not-just-for-startups/ – Lean Startup Strategy Not Just for Startups

11. Maria Popova, "Pixar Cofounder Ed Catmull on Failure and Why Fostering a Fearless Culture Is the Key to Groundbreaking Creative Work," http://www.brainpickings.org/.

CHAPTER 2

SPEED, AND WHY THE THEORY OF RELATIVITY MATTERS

1. Nicholas Carr, *The Shallows: What the Internet Is Doing to Our Brains* (New York: W. W. Norton & Company, 2011).

2. Charles H. Fine, *Clockspeed: Winning Industry Control in the Age of Temporary Advantage* (New York: Basic Books, 1998).

3. At 20 degrees Celsius, at sea level, the speed of sound is 343 meters per second, or about 768 miles per hour.

CHAPTER 3

LINEARITY IS DEAD

1. Isaac Newton, *Philosophiae Naturalis Principia Mathematica* (London: W. Dawson and Sons, 1687).

2. Not everyone agreed. The tower had been the subject of controversy, with some critics arguing that it could never be built, and many

others opposing it on artistic grounds. A group of 300 prominent artists (one member for each meter of the tower's height) objected to this tower of iron in their beautiful city: "We, writers, painters, sculptors, architects and passionate devotees of the untouched beauty of Paris, protest with all our strength, with all our indignation in the name of French taste, against the erection of this useless and monstrous Eiffel Tower." Even after the erection of the tower and its enormously positive reception by the general public, many artists remained deeply opposed to it. Guy de Maupassant, one of the most prominent writers of the time, supposedly ate lunch in the tower's restaurant every day because it was the one place in Paris from which the tower was not visible.

3. John H. Lienhard, *Inventing Modern: Growing Up with X-Rays, Skyscrapers, and Tailfins* (New York: Oxford University Press, 2003).

4. The pinnacle of this "War of the Currents" was a widely attended spectacle in Coney Island, New York, where a circus elephant named Topsy was to be executed after she was found to be too dangerous to be around people. Edison had Topsy fitted with copper-wire sandals, and before a crowd of thousands of people, an alternating current of 6,000 volts was sent through the elephant until she toppled to her side, stone dead.

5. Sadi Carnot, *Réflexions sur la Puissance Motrice du Feu: et sur les Machines Propres à Développer Cette Puissance* (Paris: Bachelier, 1824).

6. The Santa Fe Institute is probably the global focal point of thinking on the subject of complex adaptive systems. Located in New Mexico, it was founded in 1984 by a group of researchers that included Murray Gell-Mann, who won the 1969 Nobel Prize for his work on elementary particles. Gell-Mann is the researcher who found that all matter in the universe is basically made up of *quarks*, a name coined by Gell-Mann as a reference to the novel *Finnegans Wake*, by James Joyce ("Three quarks for Muster Mark!"). During the 1990s, Gell-Mann's interest turned to the emerging study of complexity, and he played a central role in the founding of the Santa Fe Institute. Most of the founders of the institute were scientists with the Los Alamos National Laboratory who felt the need for research outside the traditional disciplinary boundaries of academic departments and government agency science budgets. Their mission became to disseminate the notion of a new interdisciplinary research area called complexity theory, or the study of complex adaptive systems.

7. Based on Douglas Adams, *The Hitchhiker's Guide to the Galaxy* (Pan Books, 1979).

How the Media Discovered Networks

1. Todd G. Buchholz, *New Ideas from Dead CEOs: Lasting Lessons from the Corner Office* (New York: HarperCollins, 2007).

CHAPTER 4

INFORMATION BECOMES A FLOW

1. The highlight of the Illinois Pavilion was a stage show featuring a life-sized Audio-Animatronic figure of Abraham Lincoln. "Great Moments with Mr. Lincoln" included passages from Lincoln's speeches. Capable of more than 250,000 combinations of actions, including smiles, frowns, and gestures, the robot was voiced by actor Royal Dano.

2. Claude E. Shannon, "A Mathematical Theory of Communication," *Bell System Technical Journal* 27(3), July/October 1948, pp. 379–423.

3. Edward Thorp, *Beat the Dealer: A Winning Strategy for the Game of Twenty-One* (New York: Blaisdell Publishing, 1962).

4. Nate Silver, *The Signal and the Noise: Why So Many Predictions Fail—but Some Don't* (New York: Penguin Press, 2012).

5. Jeffrey Dean and Sanjay Ghemawat, "MapReduce: Simplified Data Processing on Large Clusters," Sixth Symposium on Operating System Design and Implementation, Seattle, WA, 2004.

6. Check out the awesome TED talk that Sean Gourley gave on exactly this topic: just Google "Sean Gourley: The Mathematics of War," and enjoy.

7. Juan Camilo Bohorquez, Sean Gourley, Alexander R. Dixon, Michael Spagat, and Neil F. Johnson, "Common Ecology Quantifies Human Insurgency," *Nature* 462, December 17, 2009, pp. 911–914.

8. Incidentally, two of the seven original bridges did not survive the bombing of Königsberg during World War II. Two other bridges were later demolished and replaced by a modern highway. The three other bridges remain, although only two of them are really from Euler's time. So, the problem of the Seven Bridges of Königsberg has now been replaced locally by the problem of the Five Bridges of Kaliningrad. For that particular problem, a Eulerian path is indeed possible, but since it must begin on one island and end on the other, it is rather impractical for tourists.

9. Jena Mcgregor, "Zappos Says Goodbye to Bosses," *Washington Post*, January 3, 2014, http://www.washingtonpost.com/.

10. http://www.zapposinsights.com/about/holacracy

11. Yannig Roth, "Crowdsourcing by World's Best Global Brands," Tiki-Toki, December 1, 2014, http://www.tiki-toki.com/.

12. Mark Granovetter, "The Strength of Weak Ties," *American Journal of Sociology* 78(6), May 1973, pp. 1360–1380.

13. John Naisbitt, *Megatrends: Ten New Directions Transforming Our Lives* (New York: Warner Books, 1982).

14. Michael Chui, James Manyika, Jacques Bughin, Richard Dobbs, Charles Roxburgh, Hugo Sarrazin, Geoffrey Sands, and Magdalena Westergren, "The Social Economy: Unlocking Value and Productivity Through Social Technologies," Mckinsey Global Institute, July 2012.

15. Accenture Technology Vision 2014.

Education in The Age of Networks

1. John Taylor Gatto, *The Underground History of American Education: A School Teacher's Intimate Investigation into the Problem of Modern Schooling* (New York: Oxford Village Press, 2001).

2. L. Rafael Reif, "Better, More Affordable Colleges Start Online: How Digital Learning Can Be a Part of Every Campus," *Time,* September 26, 2013.

3. As Daphne Koller states, "Many studies have shown that standard lecturing is not the most effective mode of instruction." The advent of the age of networks offers the opportunity to move a great deal traditional lecturing outside the classroom, and to offer an online learning format that is, in many ways, more interactive and more engaging. A study conducted in 2011 at the University of British Columbia, coauthored by physics Nobel laureate Carl Wieman, showed that students taught with the aid of a highly interactive flipped classroom approach did nearly twice as well as peers taught via traditional lectures.

4. Teresa Turiera and Susanna Cros, "Cobusiness: 50 Examples of Business Collaboration," presented By Cosociety, July 2013.

5. Although I watch most movies online, I still have the occasional urge to buy a DVD. What bothers me more than anything else, when I'm loading a DVD, is the following message on display: *WARNING: Use in other locations such as airlines, clubs, coaches, hospitals, hotels, oil rigs, prisons, schools and ships is prohibited unless expressly authorized by the copyright proprietor.* The fact that we put schools in the same league as hospitals, oil rigs, and prisons pretty much sums up how we view education.

CHAPTER 5

WHEN MARKETS STOP BEING MARKETS

1. John H. Lienhard, *Inventing Modern: Growing Up with X-Rays, Skyscrapers, and Tailfins* (New York: Oxford University Press, 2003).

2. In 1967, Kotler published the textbook *Marketing Management: Analysis, Planning, and Control*. Now in its fourteenth edition, it is the world's most widely adopted textbook in graduate schools of business. Previous marketing textbooks were highly descriptive; Kotler's was the first to draw on analytics, economic science, organizational theory, and psychology of behavior and choice. The *Financial Times* has called it one of the 50 greatest business books of all time.

3. Some additional interesting statistics:
   - 56 percent of American adults are now smartphone owners. (Source: Pew Internet & American Life Project, 2013.)
   - 75 percent of Americans bring their phones to the bathroom. (Source: Digiday, 2013.)
   - 27 percent of companies worldwide planned to implement location-based marketing in 2013. (Source: Econsultancy, 2013.)
   - Of the 70 percent of shoppers who used a mobile phone while in a retail store during the holidays, 62 percent accessed that store's site or app and only 37 percent of respondents accessed a competitor's site or app. (Source: ForeSee, 2013.)

4. David Ogilvy, "The Theory and Practice of Selling the AGA Cooker," AGA Heat Limited, 1935.

5. David Ogilvy, *Confessions of an Advertising Man* (New York: Atheneum, 1963).

6. Don Peppers and Martha Rogers, *The One to One Future: Building Relationships One Customer at a Time* (New York: Currency Doubleday, 1993).

7. David Gaughran, *Let's Get Digital: How to Self-Publish, and Why You Should* (Arriba Arriba Books, 2011).

8. September 2011 was a dire moment for Netflix. Both of its services (DVD by mail and Internet streaming) were strong, but it was clear that one technology was the future, and the other was toast. So Netflix announced that it would spin off its popular DVD business under a new brand name, Qwikster, but it was forced to reverse the decision less than a month later. The markets did not respond well, and shares of Netflix plummeted.

9. Douglas Van Praet, *Unconscious Branding: How Neuroscience Can Empower (and Inspire) Marketing* (New York: Palgrave Macmillan, 2012).

10. Timothy D. Wilson, *Strangers to Ourselves: Discovering the Adaptive Unconscious* (Cambridge, MA: Belknap Press of Harvard University Press, 2004).

11. *The Decisive Moment: How the Brain Makes Up Its Mind* by Jonah Lehrer.

12. Antonio Damasio, *Descartes' Error: Emotion, Reason, and the Human Brain* (New York: Putnam Publishing, 1994).

13. Seth Godin, *Tribes: We Need You to Lead Us* (New York: Portfolio, 2008).

14. My ideas about advertising have to a large extent been influenced by many pleasant conversations with André Duval, founder and CEO of Duval Union and one of Europe's most experienced Mad Men. More info can be found at www.duvalunion.com.

15. Barry Schwartz, *The Paradox of Choice: Why More Is Less* (New York: Ecco, 2003).

The Era of Networked Health

1. J. Watson and F. Crick, *A Structure for Deoxyribose Nucleic Acid, Nature* 171, April 1953, pp. 737–738.

2. Richard A. Epstein, *Overdose: How Excessive Government Regulation Stifles Pharmaceutical Innovation* (New Haven, CT: Yale University Press, 2006).

3. In November 2013, 23andMe was told by U.S. regulators to halt sales of its main product because it was being sold without "marketing clearance or approval." The FDA published its concern about the public health consequences of inaccurate results from the device. Clearly, the regulatory approach to the "Quantified Self" will need to adapt to a fast-changing environment.

4. Juan Enriquez, *As the Future Catches You: How Genomics and Other Forces Are Changing Your Life, Work, Health and Wealth* (New York: Crown Business, 2005).

5. Erik Topol, *The Creative Destruction of Medicine: How the Digital Revolution Will Create Better Health Care* (New York: Basic Books, 2012).

6. The offer was made in a letter to Alan Ramsay Hawley, president of the Aero Club of America, by Raymond Orteig in 1919: "Gentlemen: As a stimulus to the courageous aviators, I desire to offer, through the auspices and regulations of the Aero Club of America, a prize of $25,000 to the first aviator of any Allied Country crossing the Atlantic in one flight, from Paris to New York or New York to Paris, all other details in your care. Yours very sincerely, Raymond Orteig."

7. Scanadu is a start-up that makes medical technology devices for consumers. It was founded in 2011 by Walter De Brouwer. The company set up a lab at NASA Ames in Mountain View, where it shares space with

Singularity University. A prototype of Scanadu's first product, Scout, was unveiled on November 29, 2012. Scout is a portable electronic device that is designed to measure a number of different physiological parameters, including pulse transit time, heart rate, electrical heart activity, body temperature, heart rate variability, and blood oxygenation.

CHAPTER 6

WHEN ORGANIZATIONS BECOME NETWORKS OF INNOVATION

1. Gary Hamel, *What Matters Now: How to Win in a World of Relentless Change, Ferocious Competition, and Unstoppable Innovation* (San Francisco: Jossey-Bass, 2012).

2. Wikipedia, "Ricardo Semer," http://en.wikipedia.org/wiki/Ricardo_Semler.

3. Paper from MIT Sloan School of Management, http://ccs.mit.edu/15939/papers/assignment1/SEMCO.htm

4. Wikipedia, "Ricardo Semler."

5. Teresa Turiera and Susanna Cros, "Cobusiness: 50 Examples of Business Collaboration," presented by Co-Society, July 2013.

6. Rachel Emma Siverman, "Who's the Boss? There Isn't One," *Wall Street Journal*, June 19, 2012, http://www.wsj.com/articles/SB10001424052702303379204577474953586383604.

7. Morning Star Company, "Self-Management," http://morningstarco.com/.

8. Todd Wasserman, "Holacracy: The Hot Management Trend for 2014?," www.mashable.com.

9. "Analysis of Semco and Pixar Animated Studio as an Example of Innovative Organizations," *WritePass Journal*, December 23, 2012.

10. Henry Stewart, "8 Companies That Don't Have Managers," *Henry's Blog*, January 20, 2014, http://www.happy.co.uk/.

11. Reid Hoffman and Ben Casnocha, *The Start-Up of You: Adapt to the Future, Invest in Yourself, and Transform Your Career* (New York: Crown Business, 2012).

12. Nassim Nicholas Taleb, *Antifragile: Things That Gain from Disorder* (New York: Random House, 2012).

13. Amit Namjoshi, "Making the Organization 'Antifragile,'" February 31, 2014, http://www.managementexchange.com.

14. Jon Gertner, "The Truth About Google X: An Exclusive Look Behind the Secretive Lab's Closed Doors." *Fast Company*, May 2014, http://www.fastcompany.com/3028156/united-states-of-innovation/the-google-x-factor.

15. The story of Polaroid is as uplifting as it is sad. It is the story of Edwin Herbert Land, a true genius and a genuine entrepreneur, who was capable of capturing the hearts and minds of his customers, and who built fabulous products and created an entirely new market. But it's also a story of inevitable decline, with Land being pushed out of his own company, which eventually went bankrupt.

16. Edwin Land was the Steve Jobs of the 1950s. He was the charismatic CEO of the hottest technology start-up of its time, a maniac about product design details (he inspected every little aspect of a new camera), and a master showman on stage at his annual conventions. And a college dropout, too. No wonder Jobs liked him.

17. Johnson & Johnson Innovation Centers, http://www.jnj.com/partners/innovation-centers.

18. InnoCentive, http://www.innocentive.com/.

CHAPTER 7

CREATION AND DESTRUCTION

1. Jim Collins and Jerry I. Porras, *Built to Last: Successful Habits of Visionary Companies* (New York: HarperBusiness, 1994).

2. Phil Rosenzweig, *The Halo Effect ... and the Eight Other Business Delusions That Deceive Managers* (New York: Free Press, 2007).

3. Joseph A. Schumpeter, *Business Cycles: A Theoretical, Historical, and Statistical Analysis of the Capitalist Process* (New York: McGraw-Hill, 1939).

4. Marc Lowell Andreessen (born 1971) is an American entrepreneur, investor, software engineer, and multimillionaire. He is best known as the coauthor of Mosaic, the first widely used Web browser; as a cofounder of Netscape Communications; and as a cofounder and general partner of the Silicon Valley venture capital firm Andreessen Horowitz. He founded the software company Opsware and later sold it to Hewlett-Packard. Andreessen is also a cofounder of Ning, a company that provides a platform for social networking websites. An innovator and creator, he is one of the few people who has pioneered a software category (web browsers) used by billions of people and established several billion-dollar companies. He currently sits on the boards of Facebook, eBay, and HP, among others. A frequent keynote speaker and guest at Silicon Valley conferences, Andreessen is one of only six inductees in the World Wide Web Hall of Fame who were

announced at the first international conference on the World Wide Web in 1994.

5. Ilya Pozin, "10 Greatest Industry-Disrupting Startups of 2012," *Forbes*, July 19, 2012, http://www.forbes.com/sites/ilyapozin/2012/07/19/10-greatest-industry-disrupting-startups-of-2012.

CHAPTER 8

STRATEGY FOR THE AGE OF NETWORKS

1. Thomas Pynchon, "Entropy," *Kenyon Review* 22(2), Spring 1960, pp. 27–92.

2. Thomas Pynchon, *Gravity's Rainbow* (New York: Viking Press, 1973).

3. As an ultimate ironic tribute to his hermetic life, Pynchon has made two cameo animated appearances on *The Simpsons* playing himself, but portrayed with a brown paper bag over his head.

4. Matthew Wall, "Innovate or Die: The Stark Message for Big Business," BBC, September 4, 2014, http://www.Bbc.com/news/business–28865268.

5. Ibid.

6. Yannig Roth, "Crowdsourcing by World's Best Global Brands," Tiki-Toki, December 1, 2014, http://www.tiki-toki.com/.

7. Wall, "Innovate or Die."

8. Jessica Livingston, *Founders at Work: Stories of Startups' Early Days* (Berkeley, CA: Apress, 2007).

9. Nicholas Thomas, "11 Startups That Found Success by Changing Direction," *Mashable*, July 8, 2011.

10. Maxwell Wessel, "Why Big Companies Can't Innovate," *Harvard Business Review*, September 27, 2012.

11. Paul D. MacLean, *The Triune Brain in Evolution: Role in Paleocerebral Functions* (New York: Plenum Press, 1990).

12. Michael Hastings, "The Runaway General," *Rolling Stone*, June 22, 2010.

13. Teresa Turiera and Susanna Cros, "Cobusiness: 50 Examples of Business Collaboration," presented by Co-Society, July 2013.

14. AstraZeneca, "Opening Our Doors to Creative, Cross-Industry Partnerships," March 25, 2014, http://www.astrazeneca.com/About-Us/Features/Article/20140324—opening-our-doors-to-creative-cross-industry-partnerships.

15. Turiera and Cross, "Cobusiness."

EPILOGUE

1. Teresa Turiera and Susanna Cros, "Cobusiness: 50 Examples of Business Collaboration," presented by Cosociety, July 2013.
2. Kevin Kelly, *What Technology Wants* (New York: Penguin, 2010).
3. Douglas Adams, *The Hitchhiker's Guide to the Galaxy* (New York: Pan Books, 1979).
4. Gary Hamel, *What Matters Now: How to Win in a World of Relentless Change, Ferocious Competition, and Unstoppable Innovation* (San Francisco: Jossey-Bass, 2012).
5. John Kotter and Ken Perlman, "It's the End of an Era—Enter the Knowledgeable Networker," *Forbes*, February 13, 2013.

SOURCES

Abella, Alex. *Soldiers of Reason: The RAND Corporation and the Rise of the American Empire*. Boston: Mariner Books, 2009.

Adams, Douglas. *The Hitchhiker's Guide to the Galaxy*. New York: Pan Books, 1979.

Bohorquez, Juan Camilo, Sean Gourley, Alexander R. Dixon, Michael Spagat, and Neil F. Johnson. "Common Ecology Quantifies Human Insurgency." *Nature* 462, (December 17, 2009), pp. 911–914.

Carnot, Sadi. *Réflexions sur la Puissance Motrice du Feu et sur les Machines Propres à Développer Cette Puissance*. Paris: Bachelier, 1824.

Carr, Nicholas. *The Shallows: What the Internet Is Doing to Our Brains*.

New York: W. W. Norton & Company, 2011.

Clausewitz, Carl von. *On War*, trans. James John Graham. London: N. Trübner, 1873.

Collins, Jim, and Jerry L. Porras. *Built to Last: Successful Habits of Visionary Companies*. New York: HarperBusiness, 1994.

Damasio, Antonio. *Descartes' Error: Emotion, Reason, and the Human Brain*. New York: Putnam Publishing, 1994.

Dean, Jeffrey, and Sanjay Ghemawat. *MapReduce: Simplified Data Processing on Large Clusters*, Sixth Symposium on Operating System Design and Implementation, Seattle, WA, 2004.

Enriquez, Juan. *As the Future Catches You: How Genomics and Other Forces Are Changing Your Life, Work, Health and Wealth*. New York: Crown Business, 2005.

Epstein, Richard A. *Overdose: How Excessive Government Regulation Stifles Pharmaceutical Innovation*. New Haven, CT: Yale University Press, 2006.

Fine, Charles H. *Clockspeed: Winning Industry Control in the Age of Temporary Advantage*. New York: Basic Books, 1998.

Gatto, John Taylor. *The Underground History of American Education: A School Teacher's Intimate Investigation into the Problem of Modern Schooling*. New York: Oxford Village Press, 2001.

Gaughran, David. *Let's Get Digital: How to Self-Publish, and Why You Should*. Arriba Arriba Books, 2011.

Granovetter, Mark. "The Strength of Weak Ties," *American Journal of Sociology* 78(6), May 1973, pp. 1360–1380.

Hamel, Gary. *What Matters Now: How to Win in a World of Relentless Change, Ferocious Competition, and Unstoppable Innovation*, San Francisco: Jossey-Bass, 2012.

Hastings, Michael. "The Runaway General." *Rolling Stone*, June 22, 2010.

Hinssen, Peter. *The New Normal: Explore the Limits of the Digital World*. Ghent, Belgium: Mach Media, 2010.

Hoffman, Reid, and Ben Casnocha. *The Start-up of You: Adapt to the Future, Invest in Yourself, and Transform Your Career*. New York: Crown Business, 2012.

Kahn, Herman. *On Thermonuclear War*. (Princeton, NJ: Princeton University Press, 1960.

———— and Irwin Mann. *Techniques of Systems Analysis*. Santa Monica, CA: RAND Corporation, 1957.

———— and Anthony J. Wiener. *The Year 2000: A Framework for Speculation on the Next Thirty-Three Years*. Croton-on-Hudson, NY: Hudson Institute, 1967.

Kelly, Kevin. *What Technology Wants*. New York: Penguin, 2010.

Kotter, John, and Ken Perlman. "It's the End of an Era—Enter the Knowledge-able Networker," *Forbes*, February 13, 2013.

Lienhard, John H. *Inventing Modern: Growing Up with X-Rays, Skyscrapers, and Tailfins*. New York: Oxford University Press, 2003.

MacLean, Paul D. *The Triune Brain in Evolution: Role in Paleocerebral Functions*. New York: Plenum Press, 1990.

Naisbitt, John. *Megatrends: Ten New Directions Transforming Our Lives*. New York: Warner Books, 1982.

Newton, Isaac. *Philosophiae Naturalis Principia Mathematica*. London: W. Dawson and Sons, 1687.

Ogilvy, David. "The Theory and Practice of Selling the Aga Cooker." Aga Heat Limited, 1935.

————. *Confessions of an Advertising Man*. New York: Atheneum, 1963.

Peppers, Don, and Martha Rogers. *The One to One Future: Building Relation-ships One Customer at a Time*. New York: Currency Doubleday, 1993.

Pynchon, Thomas. "Entropy." *Kenyon Review* 22(2), Spring 1960, pp, 27–92.

————. *Gravity's Rainbow*. New York: Viking Press, 1973.

Reif, L. Rafael. "Better, More Affordable Colleges Start Online: How Digital Learning Can Be a Part of Every Campus. *Time*, September 26, 2013.

Rosenzweig, Phil. *The Halo Effect: … and the Eight Other Business Delusions That Deceive Managers.* New York: Free Press, 2007.

Schumpeter, Joseph A. *Business Cycles: A Theoretical, Historical, and Statistical Analysis of the Capitalist Process.* New York: McGraw-Hill, 1939.

Schwartz, Barry. *The Paradox of Choice: Why More Is Less.* New York: Ecco, 2003.

Shannon, Claude E. *A Mathematical Theory of Communication, Bell System Technical Journal* 27(3), July/October 1948, pp. 379–423.

Silver, Nate. *The Signal and the Noise: Why So Many Predictions Fail—but Some Don't.* New York: Penguin Press, 2012.

Taleb, Nassim Nicholas. *The Black Swan.* New York: Random House, 2007.

———. *Antifragile: Things That Gain from Disorder.* New York: Random House, 2012.

Thorp, Edward. *Beat the Dealer: A Winning Strategy for the Game of Twenty-One.* New York: Blaisdell Publishing, 1962.

Topol, Erik. *The Creative Destruction of Medicine: How the Digital Revolution Will Create Better Health Care.* New York: Basic Books, 2012.

Van Praet, Douglas. *Unconscious Branding: How Neuroscience Can Empower (and Inspire) Marketing.* New York: Palgrave Macmillan, 2012.

Watson, J., and F. Crick. "A Structure for Deoxyribose Nucleic Acid," *Nature* 171, April 1953, pp. 737–738.

Wilson, Timothy D. *Strangers to Ourselves: Discovering the Adaptive Unconscious.* Cambridge, MA: Belknap Press of Harvard University Press, 2004.

INDEX

M

ABOUT THE AUTHOR

An entrepreneur, advisor, lecturer, and writer, **Peter Hinssen** (born 1969) is one of Europe's most sought-after thought leaders on disruptive innovation. He has moved from a deep passion for all things technology to believing that digital is "merely" a spark, an enabler, and that networks are the true drivers of progress.

For more than 15 years, Peter's life revolved around technology start-ups. His first company was acquired by Alcatel-Lucent, his second was acquired by Belgacom, and his third venture (Porthus) was listed on the stock exchange in 2006. Between start-ups, he has been an entrepreneur in residence with McKinsey & Company, with a focus on digital and technology strategy.

Peter is frequently asked to lead seminars and consult on issues related to digital and technology strategy, disruptive innovation, and online tactics. He coaches business executives on developing future innovation perspectives and is a board advisor on subjects related to innovation and technology.

Peter has conducted extensive research on the adoption of technology by consumers, the impact of the networked digital society, and the fusion between business and IT.

He lectures at various business schools in Europe, such as London Business School, is a Senior Industry Fellow at the Center for Digital Transformation of the Paul Merage School of Business at University of California, Irvine, and functions as a board advisor on disruptive and digital innovation.

He has written three books on these subjects. *Business/IT Fusion*, written and published in 2008, became an instant reference for IT and business executives around the globe. In his second book, *The New Normal* (2010), Peter demonstrated how companies should explore the limits of the digital world, and what happens when technology stops being technology and just becomes "normal." His latest book, *The Network Always Wins*, explores the world beyond digital, when markets become networks and companies will also need to become networks if they are to survive.

Peter is a passionate keynote speaker who is frequently welcomed at forums and conferences around the world.

Website: www.peterhinssen.com
Twitter: @hinssen.

About Peter Hinssen's previous books,

BUSINESS/IT FUSION (2008)

Beautifully written, thoroughly researched, spiced with stories and a unique combination of humor and intellect. Peter challenges his readers to do everything they can to fuse IT and the rest of the business.

Susan Cramm
Founder and president of Valuedance, IT leadership coach, former CIO and CFO, and award-winning author of the Harvard Business Press book, "8 Things We Hate about I.T."

Read this book if you are interested in repositioning "the department previously known as IT".

Rob Goffee
Professor of Organisational Behaviour, London Business School

---

Business/IT Fusion hits the nail right on the head. It is a great mental journey on how to move beyond mere alignment thinking, and really transform IT into a true business asset.

Steve Van Wyk
Head of Operations and IT Banking, ING Group

Business/IT Fusion belongs on the bookshelf of every CIO as well as every leader who aspires to transform the IT department.

Costas Markides
Professor of Strategy and holder of the Robert P. Bauman Chair in Strategic Leadership, London Business School

THE NEW NORMAL (2010)

In *The New Normal*, Peter Hinssen presents how companies may address a society without digital limits. Quite poignantly, he points out that organizations are increasingly faced with customers and consumers who no longer tolerate limitations in terms of pricing, timing, patience, depth, privacy, convenience, intelligence.

Since the book was published, Peter toured the globe to present his views at more than 250 conferences, seminars and workshops. Time after time, audiences returned to their businesses with new insights and the will to thrive in the new normal.

"I like the ease with which you transmit such destabilizing ideas to practical people who need to act on them. I learned about some consequences of the 'new normal' that I hadn't thought about and I learned ways of conveying some that I thought about. And on top of that... I enjoyed it!"

Carlota Perez

Author of Technological Revolutions and Financial Capital: the Dynamics of Bubbles and Golden Ages.